Chaos, Fractals, and Dynamics

Computer Experiments in Mathematics

Robert L. Devaney

Department of Mathematics
Boston University
Boston, Massachusetts

♦▼ **Addison-Wesley Publishing Company**

Menlo Park, California · Reading, Massachusetts · New York
Don Mills, Ontario · Wokingham, England · Amsterdam · Bonn
Sydney · Singapore · Tokyo · Madrid · San Juan

For my mother

Cover:
The cover image shows a computer graphics display of a small copy of the Mandelbrot set. (Chapter 8)

This book is published by the Addison-Wesley Innovative Division.

Apple is a registered trademark of Apple Computer, Inc.
Macintosh is a trademark of Apple Computer, Inc.
IBM is a registered trademark of International Business Machines Corporation.

ISBN 0-201-23288-X

7 8 9 10 - ML - 95

Preface

There are a number of reasons why I have written this book. By far the most important is my conviction that we in mathematics education — whether on the secondary or collegiate level — fail miserably to communicate the vitality of contemporary mathematics to our students. Let's face it: most high school and college mathematics courses emphasize centuries-old mathematics. As a consequence, many students think that new mathematics ended in the time of Euclid and Pythagoras (or, if they know calculus, in the time of Newton and Leibniz). Many students feel that mathematics is a dead discipline, that, as a career, it leads nowhere but to teaching (and thereby the perpetuation of a dead discipline). They have no appreciation or understanding of what a mathematician really does or thinks about.

I can only imagine the outcry that would occur if a physicist, chemist, or biologist taught only seventeenth century science in their secondary school courses. Yet we as mathematicians think nothing of doing exactly that! It is true that mathematics is a "cumulative" science, but there are exceptions to this and we should strive at every opportunity to give our students a glimpse of what is new and exciting in mathematics.

I feel that one such opportunity is presented by the field of dynamical systems, the branch of mathematics that studies processes that move or change in time. Dynamical systems are encountered in all branches of science. For instance, the changing weather patterns in meteorology, the ups and downs of the stock market in economics, the growth and decline of populations in ecology, and the motion of the planets and galaxies in astronomy are all examples of dynamical systems. The field of dynamical systems is an important area of contemporary research in mathematics which enjoys the additional advantage of being quite accessible to the nonmathematician. One aspect of dynamical systems involves iterating a mathematical function over and over, and then asking what happens. Amazingly, for even the simplest mathematical expressions such as quadratic or cubic functions of a real or complex variable, nobody knows the complete answer. Indeed, there are many mathematicians worldwide working on precisely this question. Note how easy it is to explain this unsolved problem to high school students. With the quadratic formula and elementary graphing techniques, these students feel they know *everything* there is to know about quadratic functions. My experience is that they are quite surprised to know that not everything is known, and then many are quite eager to plunge in and explore on their own.

Since computers are readily available to most students, now is an especially auspicious time to introduce students to dynamical systems, and this is my second reason for writing this book. The process of iteration is impossible to carry out by hand but extremely easy to carry out with a calculator or computer. A very simple six- or seven-line BASIC program allows a student to compute hundreds and thousands of iterations of a single function. Students get the feeling that they have the power to explore the uncharted wilderness of the dynamics of quadratic functions, not to mention the many, many other simple functions whose dynamics are less well understood. This is a radical new development in mathematics instruction. It gives mathematics an experimental component, a laboratory. Much as the physicists, chemists, and biologists have long used the laboratory as an essential component of their introductory courses, now we in mathematics have the same opportunity, and the results should be a much higher appreciation for and recognition of the importance of research mathematics by contemporary students.

My last reason for writing this book is to expose students to the great beauty of the field of dynamical systems. We in mathematics continually tell our students how beautiful mathematics is; the order and the structure and the interrelations between the various branches of mathematics which attract us as mathematicians should motivate us to attempt to communicate this to others as well. But, more often than not, we fail to show our students the beauty that we see. However, now, with the computer as a tool, this beauty is at our fingertips. The ease with which students can compute and display extraordinarily intricate Julia sets and the Mandelbrot set make these geometric images quite accessible to them. My experience is that students are quite enthralled by these images. They readily produce Julia set after Julia set and then begin to ask, "What is going on here?" This is the beginning of a novel mathematical experience, the realization that, in many cases, nobody knows the answer to this question.

It seems to me, then, that the accessibility and allure of dynamical systems theory make it a natural subject to introduce to high school and college students. I feel that the exposure to contemporary mathematics is much, much more valuable in the long run than a continued push to learn more tools and techniques for "what comes later." I also feel that the realization that mathematics is an alive and vital discipline will help dispel the thoughts of students that mathematics as a discipline leads nowhere.

A NOTE TO TEACHERS

The goal of this book is to provide teachers with material that will enable them to excite their high school or college students about mathematics. The mathematical exposition should be comprehensible to students with a solid Algebra II background. Some trigonometry is necessary for later chapters (particularly Chapters 7 and 11) and the sine and cosine functions are used as occasional examples throughout. However, a detailed knowledge of trigonometry is not necessary to read most of the book.

This book can be used in a classroom setting in a variety of ways. The material in this book can provide "special topics" that may be sprinkled about various areas of the secondary school or college curriculum. For example, the material on iteration and graphical analysis is a topic that fits naturally into a discussion of functions and graphing techniques in general. The computer programs to generate orbits, fractals, Julia sets, and the like are quite short and can easily be incorporated into an elementary programming course. Many of the topics in the book make excellent projects for mathematics clubs. The ideas of dynamics provide a perfect topic for a one-semester course in the final year of high school (replacing, perhaps, an oversimplified calculus course) or as a course in college used by liberal arts students to fulfill a mathematics requirement. Such courses might even inspire some students to continue taking mathematics rather than jeopardize future possibilities for a career in science by discontinuing their mathematical studies too early.

In each chapter at least one new purely mathematical topic is introduced. These range from the concepts of iteration, orbits, graphical analysis, and bifurcation in the earlier chapters, to the algebra of complex numbers, polar representations, and complex square roots in later sections. A number of opportunities for the reader to pursue further work in this area are also included. These are classified as exercises, experiments, and projects. Exercises usually involve routine computations using material just introduced. Experiments require the reader to use computer programs from the text to "discover" related phenomena. Projects require the reader to modify these programs in various ways to produce different output or to make the computer program run faster.

A few words about the programs: The computer programs presented in this text were written in the BASIC language. The rationale for this is to make the material in the book accessible to as wide an audience as possible. Students or instructors who know or have access to other languages such as PASCAL or C should have no difficulty converting the programs to these languages. Indeed, this would be desirable, since many of the programs involving Julia and Mandelbrot sets take quite a while to run on a personal computer.

This book is not intended to be an introductory programming text; however, with a little coaching, students who have no programming experience can use the book. Only the statements and graphics commands that are absolutely necessary for our purposes are introduced. Students with some programming experience should be able to build on the programs in the text to create custom-made programs, which include additional features such as color output or graphics input using a mouse. On the other hand, students with no programming experience can simply type the given programs into the computer and then observe the dynamical behavior. It should be noted that the early programs in the text are quite short, often less than ten lines long. As students gradually become comfortable using the computer, they can make the simple modifications to the given programs suggested in the projects and thereby become more adept at programming.

Because the major projects in the book demand so much computer time, students often enjoy starting their programs after school, running them in the evening, and then returning to find the output early the next morning.

This book has been intentionally designed so that the level of difficulty and sophistication required increases as it progresses. In the beginning, students use only a scientific calculator or a simple six-line BASIC program to calculate orbits. The only mathematics necessary are the concepts of a function and its graph. The heart of the book is the chapters on Julia sets, the Mandelbrot set, and fractals. Here the student will be exposed to somewhat longer graphics programs as well as geometric structures in the complex plane. The final section on the Julia sets of transcendental functions should be regarded as an optional special "treat" for those who wish to immerse themselves fully in the theory of Julia sets. This section introduces complex functions such as the exponential, sine, and cosine functions, so the mathematical going gets rougher at times.

The basic goal of the book is to introduce the concept of iteration and to investigate and explain the wonderful structures — Julia sets, the Mandelbrot set, and fractals — that result from iteration. Accordingly, Chapters 6 and 7 (Julia sets), 8 (the Mandelbrot set), and 9 (fractals) form the heart of the book. Chapter 9 is independent of Chapters 6-8 and can be read before these sections. Also, Chapter 7 merely gives additional algorithms for computing Julia sets and so may be skipped.

ACKNOWLEDGMENTS

It is a pleasure to acknowledge the support and assistance of a number of individuals while this book was being written. Jon Choate, Patricia S. Davidson, Paula Drewniany, Paul Goodey, Alan Hoffer, Phil Straffin, Julia Swan, and Lois Voss read and made helpful comments on portions of the manuscript. Chris Allgyer, George Best, and Connie Overzet Wheeler made extensive suggestions on earlier drafts of the text. Harvey Keynes and Tom Berger provided me with a forum at the University of Minnesota to try out these materials with groups of high school teachers and students. I am especially indebted to Tom Scavo who pointed out a number of inaccuracies in the first printing of the book. I would also like to thank a number of colleagues and students who wrote the programs that generated the computer graphics for this book, including Paul Blanchard, Elwood Devaney, Stefen Fangmeier, Chris Mayberry, Chris Small, Sherry Smith, Craig Upson, Gert Vegter, and especially Scott Sutherland. Finally, this book was written with partial support from the National Science Foundation. It is a pleasure to thank Richard Millman and John Thorpe for their encouragement at the outset of this project.

Robert L. Devaney
Boston, Mass.
March, 1989

Contents

Chapter 0. A Mathematical Tour 1

Chapter 1. Iteration 7
    1.1 Mathematical Functions 7
    1.2 Iteration Using a Scientific Calculator 8
    1.3 Functional Notation 10
    1.4 Orbits 12
    1.5 Fixed and Periodic Points 17
    1.6 An Application from Ecology 19

Chapter 2. Iteration Using the Computer 23
    2.1 The Program ITERATE1 23
    2.2 The Logistic Function 26
    2.3 Computer Graphics 31

Chapter 3. Graphical Analysis 39
    3.1 The Graph of a Function 39
    3.2 Using Graphical Analysis 40
    3.3 Attracting and Repelling Fixed Points 45
    3.4 Stable and Unstable Orbits 50
    3.5 Attracting and Repelling Periodic Points 51
    3.6 Higher Iterates 52

Chapter 4. The Quadratic Family 57
    4.1 Escaping Orbits of the Quadratic Function 57
    4.2 The Interesting Orbits 61
    4.3 The Period-Doubling Bifurcation 63
    4.4 The Chaotic Quadratic Function 68
    4.5 The Orbit Diagram 70

Chapter 5. Iteration in the Complex Plane 75
    5.1 Complex Numbers 75
    5.2 The Program ITERATE4 78
    5.3 The Julia Set 83

Chapter 6. The Julia Set: Basin Boundaries 85
    6.1 Escaping Orbits 86
    6.2 The Program JULIA1 87
    6.3 Magnifying the Julia Set 90

Chapter 7. The Julia Set: Other Algorithms 97
    7.1 Polar Representation of a Complex Number 97
    7.2 The Squaring Function Again 101
    7.3 The Program JULIA2 102
    7.4 Fractal Dust 105
    7.5 The Boundary-Scanning Method 111

Chapter 8. The Mandelbrot set 113
    8.1 Critical Points and Orbits 113
    8.2 Construction of the Mandelbrot Set 115
    8.3 The Program MANDELBROT1 116
    8.4 Refinements of MANDELBROT1 120
    8.5 What the Mandelbrot Set Means 124

Chapter 9. Geometric Iteration: Fractals 129
    9.1 Fractals 129
    9.2 The Sierpinski Triangle 130
    9.3 The Cantor Set 133
    9.4 The Koch Snowflake 137
    9.5 Computing Fractals 139
    9.6 Fractional Dimension 144
    9.7 Fractals and Dynamics 147

Chapter 10. Chaos 151
    10.1 The Squaring Function Again 152
    10.2 Sensitive Dependence 154

Chapter 11. Julia Sets of Other Functions 159
    11.1 Higher-Degree Polynomials 159
    11.2 Euler's Formula 165
    11.3 Julia Sets of Transcendental Functions 167
    11.4 Exploding Julia Sets 172

For Further Reading 177
Index 179

Chapter 0

A Mathematical Tour

This book deals with some very interesting, exciting, and beautiful topics in mathematics — topics which, in many cases, have been discovered only in the last decade.

In this book you will meet unfamiliar terms such as iteration, orbits, chaos, fractal, Julia sets, the Mandelbrot set, and many more. Topics such as these are too new to have entered the high school or college mathematics curriculum, but at the same time they are so stimulating and alluring that students considering a background in any area of science should be exposed to them. We would go even further — some of the images that result from these topics are so captivating that they may even be called art and should be experienced by any educated person.

This section is called Chapter 0 for a reason: Unlike the remaining chapters, we will do no mathematics in this section. Rather, we will take you on a tour of some of the remarkable images that will come later. We won't explain what these images mean now, or even how to generate them. All we will do is whet your appetite for what comes later when you will see that the mathematics that lies behind these images is even prettier than the pictures themselves.

The subject of this book is *dynamical systems,* the branch of mathematics that attempts to understand processes in motion. Such processes occur in all branches of science. For example, the motion of the stars and the galaxies in the heavens is a dynamical system, one that has been studied for centuries by thousands of scientists. The ups and downs of the stock market is another system that changes in time, as is the weather throughout the world. The changes chemicals undergo, the rise and fall of populations, and the motion of a simple pendulum are classical examples of dynamical systems in chemistry, biology, and physics. Clearly, dynamical systems abound.

What does a scientist wish to do with a dynamical system? Well, since the system is moving or changing in time, the scientist would like to predict where the system is heading, where it will ultimately go. Will the stock market go up or down? Will it be rainy or sunny tomorrow? Will these two chemicals explode if they are mixed in a test tube?

Clearly, some dynamical systems are predictable, whereas others are not. You know that the sun will rise tomorrow and that, when you add cream to a cup of coffee, the resulting "chemical" reaction will not be an explosion. On the other hand, predicting the weather a month from now or the Dow Jones average a week from now seems impossible. You say, "I know why I can predict the motion of the planets and simple chemical reactions but not the behavior of the weather or the economy — there are simply too many variables present in meteorological or economic systems to make long-term prediction possible!" This is indeed true in these cases, but this is by no means the complete answer. One of the remarkable discoveries of twentieth century science is that very simple systems, even systems depending on only one variable, may behave just as unpredictably as the stock market, just as wildly as a turbulent waterfall, and just as violently as a hurricane. The culprit, the reason for this unpredictable behavior, has been called "chaos" by mathematicians.

Because chaos has been found to occur in the simplest of systems, scientists may now begin to study unpredictability in its most basic form. It is to be hoped that the study of simple systems will eventually allow scientists to find the key to understanding the turbulent behavior of systems involving many variables such as weather or economic systems.

In this book we discuss chaos in its simplest possible setting. We will see that chaos occurs in elementary mathematical objects — objects as familiar as quadratic functions — when they are regarded as dynamical systems. Now you may object because you feel that you know all there is to know about quadratic functions — after all, they are easy to evaluate and to graph. But the key words here are dynamical systems. We will treat simple mathematical operations like taking the square root, squaring, or cubing as a dynamical system by repeating the procedure over and over. This process is called *iteration*. We will perform the same mathematical operation countless times using the output of the previous operation as the input for the next. What we will find is, in many cases, chaos, or unpredictability, or extremely complicated results. So this is one way to generate and then to begin to understand chaos, with simple mathematical models. The operation of iteration can be carried out quite easily on a computer and the results tabulated and viewed using computer graphics. Since these simple expressions when iterated lead to complete unpredictability, there is little wonder why much more complicated systems lead to this too.

Our goal will be to take these simple mathematical expressions, iterate them, and see what happens. Sometimes we will find that, when we input certain numbers into the process, the result is completely predictable, while other numbers yield results that are often bizarre and totally unpredictable. For the types of expressions we will consider, the set of numbers that yield chaotic or unpredictable behavior is called the Julia set after the French mathematician Gaston Julia, who first formulated many of the properties of these sets in the 1920s.

These Julia sets are spectacularly complicated, even for quadratic functions. They are examples of fractals. These are sets which, when magnified over and over again, always resemble the original image. The closer you look at a fractal, the more you see exactly the same object. Moreover, fractals naturally have a dimension that is not an integer, not 1, not 2, but often somewhere in between, such as dimension 1.4176, whatever that means!

Here are some examples. In Plate 1 we show the Julia set of the simple mathematical expression $z^2 - 1$, where z is a complex number. As we will describe later, the black points you see in this picture are by no means chaotic. They are points which, under iteration of the expression $z^2 - 1$, eventually tend to cycle back and forth between 0 and -1. This is not at all apparent right now, but by the time you have read Chapter 6, you will consider this example a good friend. Points that are colored in this picture also behave predictably: They are points that escape, that tend to infinity under iteration. The colors here simply tell us how quickly a point escapes. The boundary between these two types of behavior — the interface between the escaping and the cycling points — is the Julia set. Look closely at this boundary. It looks basically like a large ball decorated with many smaller balls. If we magnify a portion of this picture, as we have in Plate 2, you see that each of these smaller balls is decorated with many many more tinier balls. In Plate 3 we magnify more so that you see that this process continues, the smaller decorations are in turn decorated by even smaller balls, and so forth. This is the concept of a fractal, one of the central notions in this book.

Here are some other Julia sets for quadratic functions and their magnifications. Plate 4 shows the Julia set known as Douady's rabbit, the Julia set of $z^2 - .122 + .745i$. Plate 5 magnifies the rabbit and shows that this rabbit has ears everywhere! Plates 6 and 7 show another Julia set of a quadratic function, this time $z^2 + .360284 + .100376i$. These Julia sets possess an amazing amount of complexity.

All these Julia sets correspond to mathematical expressions that are of the form $z^2 + c$. Here c is a complex number and, as we see, as c varies, these Julia sets change considerably in shape. How do we understand the totality of all of these shapes, the collection of all possible Julia sets for quadratic functions? The answer is called the Mandelbrot set. The Mandelbrot set, as we will see in Chapter 8, is a dictionary, or picture book, of all possible quadratic Julia sets. It is a picture in the c-plane that provides us with a road map of all possible quadratic Julia sets. This image, first viewed in the late 1970s by Mandelbrot and others, is quite important in dynamics. It completely characterizes the Julia sets of quadratic functions. It has been called one of the most intricate and most beautiful objects in mathematics.

Plate 8 shows the full Mandelbrot set. Note that it consists of a basic central cardioid shape, with smaller balls attached. Unlike the Julia set of $z^2 - 1$, these decorations seem to have antennae attached. Plates 9 and 10 show some of these decorations with their antennae. In Plates 11 and 12 we see that some of these antennae become quite complicated — they even resemble seahorses at times or groups of elephants marching around a curve. Magnifications of these images (Plates 13, 14) show the incredible complexity of the Mandelbrot set. Plates 15–17 yield a surprise; buried deep within the Mandelbrot set are smaller copies of the entire set. So we can play the same game over and over, delving deeper and deeper into the Mandelbrot set, finding more and more interesting phenomena.

The Julia sets of complex functions come in all sorts of different shapes. Plate 18 depicts the Julia set of the expression $\pi i e^z$, while Plate 19 shows the corresponding image of $\pi i \tan z$.

Julia sets are always fractals, as Plates 20–22 show. These plates depict the Julia set of the trigonometric sine function. Basically, the black, or "stable," region looks like a collection of infinitely many snowmen. But if we magnify a portion of a snowman, we see that its arms have infinitely many joints (Plate 21) and each arm has infinitely many pimples (Plate 22).

Julia sets may behave quite strangely when their defining parameters change. For example, they may literally explode as parameters vary. Look at the large black region for $\sin z$ in Plate 20, but watch what happens to it when we consider $(1 + .2i)\sin z$ instead (Plate 23). Plate 24 depicts the Julia set of $0.36e^z$, with its very large black region, but Plate 25 shows how this nonchaotic region evaporates when we consider instead $0.38e^z$. There are many examples of this type of behavior, as Plates 26–29 further illustrate. Plate 26 shows a large black region for the expression $(.6 + .8i) \sin z$, while the remaining three images show how this region disappears when we consider $(.61 + .81i) \sin z$ instead.

It is quite interesting to watch Julia sets as parameters vary continuously. For example, in Plates 30–36, we consider the Julia sets for the functions $c \cos z$ as c decreases from π to 2.94. Plate 30 shows a big black stable region for $\pi \cos z$, together with infinitely many satellite black regions. For $2.97 \cos z$, this black region has contracted to a cauliflower-shaped region (Plate 31) and just below 2.97, this region explodes in color (Plate 32). For $2.96 \cos z$ (Plate 33) we note that the smaller black bubbles appear to be exploding as well, and further magnification for $2.955 \cos z$ (Plate 34) shows that, indeed, the regions inside the original figure bear a striking resemblance to the original figure. Plates 35–36 show fine detail of these Julia sets for $2.95 \cos z$ and $2.94 \cos z$.

The images in this mathematical tour show quite clearly the great beauty of mathematical dynamical systems theory. But what do these pictures mean and how are they produced? These are questions that we will answer in the remainder of this book.

Color Plates

Plates 1–3. The Julia set of $z^2 - 1$ and several magnifications. These plates illustrate the fractal, yet very regular, nature of Julia sets. Decorations on top of decorations on top of decorations....

Plates 4, 5. Douady's rabbit: the Julia set of $z^2 - .122 + .745i$. Again we see the fractal nature of a Julia set: everywhere we look, the rabbit's ears keep popping up.

Plates 6, 7. More quadratic Julia sets: this time the Julia set of $z^2 + .360284 + .100376i$. As we vary the constant c in the expression $z^2 + c$, we seem to find completely different structures for the Julia set.

Plate 8. The Mandelbrot set. Incredibly intricate, this set has been called the most complicated yet most beautiful object in mathematics.

Plates 9, 10. The Mandelbrot set is decorated with infinitely many "balls" with "antennae."

Plates 11, 12. Some of the decorations on the Mandelbrot set look like monsters; others look like chains of elephants.

Plates 13, 14. Further magnification of portions of the Mandelbrot set show the rich variety of shapes that make up the set.

Plates 15–17. There are "baby" Mandelbrot sets everywhere in the Mandelbrot set. Here we find several different copies at a magnification of 10^4.

Plate 18. The tangled and twisting strings in the Julia set of $\pi i e^z$.

Plate 19. Fractal footsteps on the beach: the Julia set of $\pi i \tan z$.

Plates 20–22. An infinite snowman: the Julia set of $\sin z$ and magnifications.

Plate 23. An explosion into chaos: the Julia set of $(1 + 0.1i) \sin z$.

Plates 24, 25. Spiralling galaxies emanating from a fractal fountain: the Julia sets of $(1/e) \exp z$ and $(1/e + 0.1) \exp z$.

Plates 26–29. A Siegel Disk crumbles: the Julia sets of $(.6 + .8i) \sin z$ and its neighbors.

Plates 30–36. A chain of explosions in the cosine family: the Julia sets of $c \cos z$ as c decreases from π to 2.94.

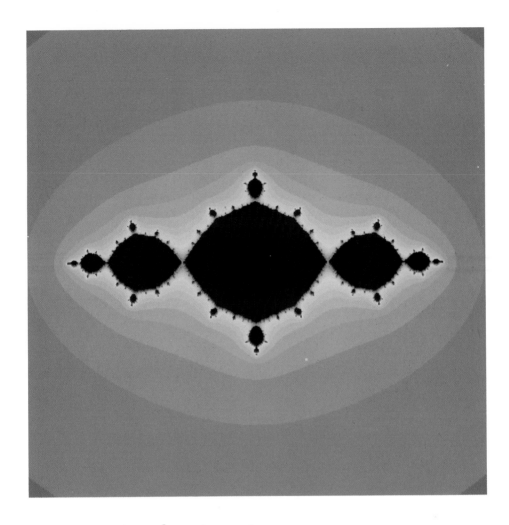

Plates 1–3. Julia set of z^2-1 and magnifications.

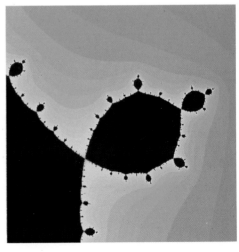

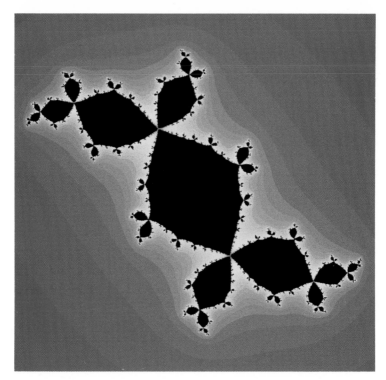

Plate 4. *The Rabbit.*

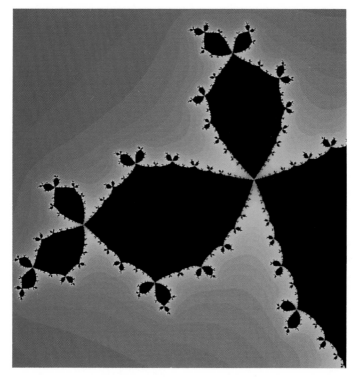

Plate 5. Magnification of Plate 4.

Plate 6. *The Dragon.*

Plate 7. Magnification of Plate 6.

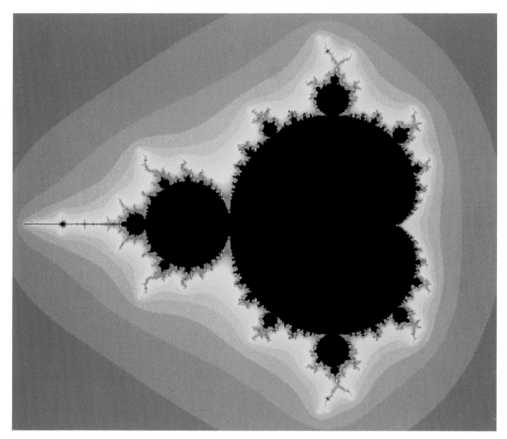

Plate 8. The Mandelbrot set.

Plates 9, 10. Details of the Mandelbrot set.

Plate 11. *Elephants.*

Plate 12. *Monsters.*

Plates 13, 14. Magnifications of Plate 12.

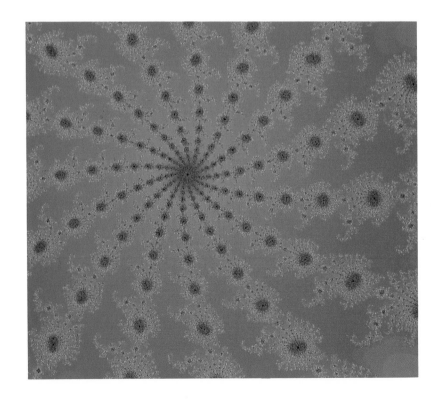

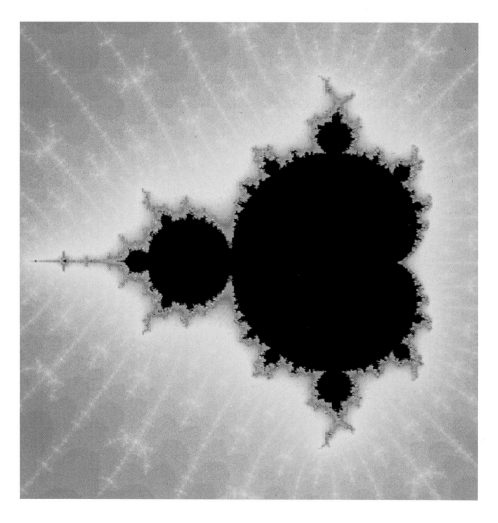

Plates 15–17. Small copies of the Mandelbrot set within $\mathcal{M}$.

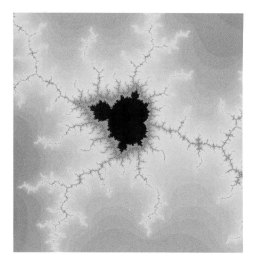
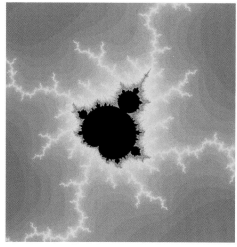

Plate 18. The Julia set of $\pi i e^{z}$.

Plate 19. The Julia set of $\pi i \tan z$.

Plates 20–22. Infinite snowmen: the Julia set of sin z and magnifications.

Plate 23. The Julia set for $(1 + 0.1i)\sin z$.

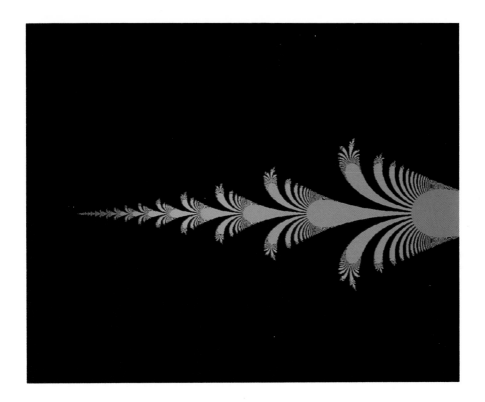

Plates 24, 25. The exploding exponential.

Plates 26–29. Collapse of a Siegel disk.

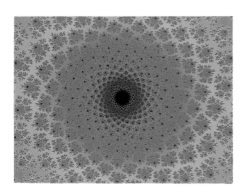
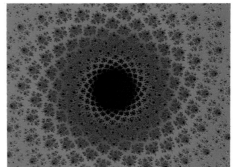

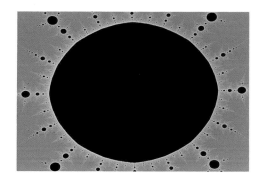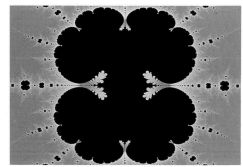

Plates 30–32. Julia sets of $\pi \cos z$, $2.97 \cos z$, and $2.965 \cos z$.

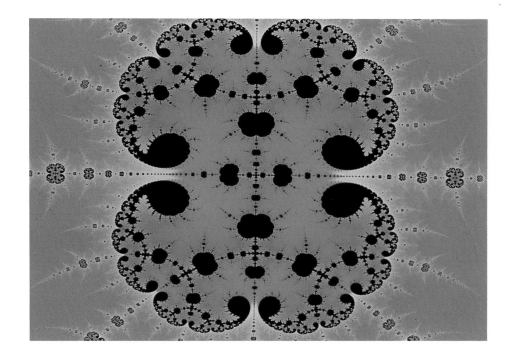

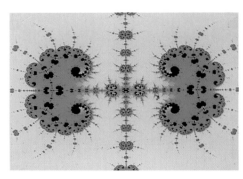

Plates 33–36. Julia sets for $2.96 \cos z - 2.94 \cos z$.

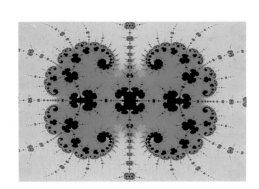
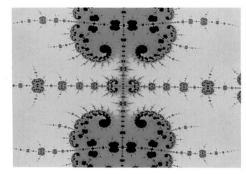

Chapter 1

Iteration

In this chapter we introduce the basic operation of dynamical systems, iteration. To iterate means to repeat a process over and over again. In dynamics, the process that is repeated is the evaluation of a mathematical function, although we will see later that many other processes may be iterated as well. Without being too technical for the moment, let's begin by describing what we mean by a *function*.

1.1   Mathematical Functions

A function is an *operation* that converts certain numbers into other, possibly different numbers. We stress here the word operation, for it is important to think of a function as an action or a process that changes numbers each time that it is applied or invoked.

A good example is the square root *function*. Taking the square root of a nonnegative number is the process that converts this number to a new number, its nonnegative square root. For example,

$$\sqrt{4} = 2$$
$$\sqrt{16} = 4$$
$$\sqrt{2} = 1.41421356\ldots$$

and so forth. We think of the symbol $\sqrt{}$ as the function that converts certain numbers — the inputs — into other numbers — the outputs.

All functions work in essentially the same way. We are given a collection of numbers — the inputs — and an operation that converts these numbers to other numbers — the outputs.

Another familiar example of a function is the squaring function. Given any number x, we may compute with ease the product of that number with itself, or x^2. Note that any real number may serve as input to the squaring function, but the output is always a nonnegative number. Thus the action of squaring gives us another example of a mathematical function.

One very important property of a function is that, whenever we apply a function to a given input, we get one and only one answer as a result. For example, the square root function yields one and only one output, the nonnegative square root, whenever it is invoked. Similarly, squaring yields one and only one result for any given input.

One of the best ways to illustrate what a function does is to use a scientific calculator. A scientific calculator often comes equipped with a variety of "function" keys — buttons that play the role of mathematical functions. You have undoubtedly seen such a calculator with buttons labeled "sin," "cos," "exp," and so forth. Pressing these buttons computes the important mathematical functions called the sine, the cosine, and the exponential. You need not know what the sine, cosine, or exponential functions are to read this book, but we will often use these functions as examples, so you should have a calculator or computer at your disposal that will enable you to use these functions.

Now let's turn to iteration, the process of evaluating a function repeatedly.

1.2   Iteration Using a Scientific Calculator

As we just mentioned, most scientific calculators have a number of special keys that correspond to important functions such as x^2, $\sqrt{x}$, $\sin x$, $\cos x$, and others. Depressing one of these keys after an initial number x has been input computes the value of the corresponding function. Iteration involves repeating this process over and over, using the result of the previous computation as the input for the next. That is, the process of iteration consists of selecting an initial x as seed or input and then striking a particular function key repeatedly.

For example, to iterate the square root function, all we need do is select an initial x–value and depress the $\sqrt{x}$ key several times. If we select the initial value $x = 256$, then we compute in order

$$\sqrt{256} = 16$$
$$\sqrt{16} = 4$$
$$\sqrt{4} = 2$$
$$\sqrt{2} = 1.414214...$$
$$\sqrt{1.414214...} = 1.189207...$$
$$\sqrt{1.189207...} = 1.090508...$$
$$\sqrt{1.090508...} = 1.044274...$$
$$\vdots$$

Continuing in this fashion, we see that repeated applications of the square root function eventually yield the number 1, which then remains unchanged or fixed under subsequent iterations.

This result occurs no matter which positive x we use as initial seed. For example, if we select $x = .5$, then we find

$$\sqrt{.5} = .707107\ldots$$
$$\sqrt{.707107\ldots} = .840896\ldots$$
$$\sqrt{.840896\ldots} = .917004\ldots$$
$$\sqrt{.917004\ldots} = .957603\ldots$$
$$\sqrt{.957603\ldots} = .978572\ldots$$
$$\sqrt{.978572\ldots} = .989228\ldots$$
$$\sqrt{.989228\ldots} = .994599\ldots$$
$$\vdots$$

You should experiment with several other initial positive x values to check that the values displayed always tend to 1. One of our main goals will be to explain why this happens.

Experiment 1.1 Use your calculator to check that iteration of the square root function always eventually leads to the number 1 being displayed, as long as the initial input is a positive number. Try, as initial inputs, $x = 10$, $x = 0.1$, $x = 123,456$, and $x = 0.123456$.

1.3   Functional Notation

At this point, let us introduce some mathematical notation that will be useful as we proceed. We will denote a mathematical function such as the square root function by

$$S(x) = \sqrt{x}$$

or just S, for short. Here, S means the operation of computing the nonnegative square root of the nonnegative number x. Remember, we always think of a function as an operation or a process that we apply to given numbers. When we replace the x with a particular nonnegative number, then the value of the function becomes a particular number too. For example,

$$S(256) = 16$$
$$S(9) = 3$$
$$S(1) = 1$$

So S gives us a rule for converting certain initial inputs x into new numbers, the result of applying the square root function, $S(x)$.

This notation may appear strange and cumbersome at first, but we will see that it is quite handy later. We will deal with many different functions in this text, for example, the squaring function

$$T(x) = x^2$$

the cosine function

$$C(x) = \cos x$$

and the exponential function

$$E(x) = \exp x$$

The notation $F(x)$ for a function is particularly useful for iteration, for we can apply the function F any number of times in succession. That is, we can first compute $F(x)$. Since $F(x)$ is itself a number, we can apply F to it, thereby getting the new value $F(F(x))$. Similarly, we may apply F a third time, getting $F(F(F(x)))$. Clearly, we need better notation for successive applications of a function F, for otherwise we'll get lost among all the parentheses. So we will write $F^j(x)$ to mean the jth application of the function F. That is,

$$F^1(x) = F(x)$$
$$F^2(x) = F(F(x))$$
$$F^3(x) = F(F(F(x)))$$

For example, if $S(x) = \sqrt{x}$, then

$$S(x) = \sqrt{x}$$
$$S^2(x) = \sqrt{\sqrt{x}}$$
$$S^3(x) = \sqrt{\sqrt{\sqrt{x}}}$$

and so forth.

You should always remember that $F^2(x)$ does not mean the square of the number $F(x)$, or $F(x) \cdot F(x)$. Rather, $F^2(x)$ means the second iterate of F at x, namely, $F(F(x))$. There is a big difference between these two operations!

Returning to the square root example, if we let $S(x) = \sqrt{x}$ and again choose $x = 256$ as our initial input, then we have:

$$S(256) = 16$$
$$S^2(256) = S(16) = 4$$
$$S^3(256) = S(4) = 2$$
$$S^4(256) = S(2) = 1.414214\ldots$$
$$S^5(256) = 1.189207\ldots$$
$$S^6(256) = 1.090508\ldots$$
$$S^7(256) = 1.044274\ldots$$
$$\vdots$$
$$S^{20}(256) = 1.000005\ldots$$

Again we note that successive iterates of 256 converge quite quickly to 1. As we observed above, this happens no matter which positive number is initially input into the calculator. For example, if the initial input is 200, we get

$$S(200) = 14.14214\ldots$$
$$S^2(200) = 3.760603\ldots$$
$$S^3(200) = 1.939227\ldots$$
$$S^4(200) = 1.392561\ldots$$
$$\vdots$$
$$S^{20}(200) = 1.000005\ldots$$

Even if the initial input x satisfies $0 < x < 1$, we still find that iterates tend to 1:

$$S(.2) = .4472136\ldots$$
$$S^2(.2) = .6687403\ldots$$
$$S^3(.2) = .8177654\ldots$$
$$\vdots$$
$$S^{20}(.2) = .9999985\ldots$$

1.4   Orbits

Let's iterate more of these functions. For example, what happens if we repeatedly strike the x^2 key? Clearly, if $x > 1$, repeated squaring tends to enlarge the results. In fact, after only a few iterations, repeated squaring leads to an overflow message from the calculator. This happens because the numbers have become too large for the calculator. If we write $T(x) = x^2$, another way of saying this is

$$T^n(x) \to \infty \quad \text{as} \quad n \to \infty$$

if $x > 1$; the notation $T^n(x) \to \infty$ means that the nth iterate of any $x > 1$ tends to get larger and larger, eventually becoming larger than any positive number whatsoever. In plain English, we say that $T^n(x)$ approaches infinity as n tends to infinity.

What if $0 < x < 1$? Then iteration of T yields a different answer. Successive squarings of such numbers yield smaller and smaller positive results, so

$$T^n(x) \to 0 \quad \text{as} \quad n \to \infty$$

when $0 < x < 1$. This means that $T^n(x)$ gets closer and closer to 0 as n increases. Of course, $T^n(x)$ never equals 0 exactly for any n, because the only number whose square is 0 is 0 itself; the iterates of x simply approach 0 without ever reaching 0. It is true that your calculator will eventually display the number 0.0000... when you compute these iterates, but this simply means that these numbers have become too small to be "seen" on the calculator's display. Finally, in the intermediate case, $x = 1$, it is clear that $T^n(x) = 1$ for all n. This point is called a *fixed point* because T leaves it fixed; the point never changes, or "moves," under iteration of T. So the iteration of T yields three different behaviors, depending upon whether $0 < x < 1$, $x = 1$, or $x > 1$.

Exercise 1.2 Extend this analysis to the case of negative x-values. What can you say about $T^n(x)$ if (a) $x < -1$; (b) $-1 < x < 0$; (c) $x = -1$?

The list of successive iterates of a point or number is called the *orbit* of that point. For example, the orbit of 2 under the squaring function $T(x) = x^2$ is given by the list:

$$2, 4, 16, 256, 65536, \ldots$$

Similarly, the orbit of .5 is

$$.5, .25, .0625, .00390625, \ldots$$

The list of numbers that make up an orbit may not resemble any "orbit" that you have ever seen, such as the orbit of a spaceship or the orbit of a planet, but there is a connection. Iteration of functions is intimately associated with the field of mathematics known as differential equations. This field uses calculus to study the behavior of processes in motion. For example, differential equations enabled Newton to formulate his theory of gravitational attraction, which in turn allowed him to predict the motion, or orbits, of heavenly bodies. Our simple iteration process also describes a system in motion, and we therefore adopt similar terminology.

Let's use the calculator to compute some other orbits. For example, let us use the $\sin x$ button on the calculator to compute the orbit (in radians) of any initial input to the sine function. If we let

$$S(x) = \sin x$$

and choose the initial value $x = 1.57$, we find

$$S(1.57) = .999\ldots$$
$$S^2(1.57) = .841\ldots$$
$$S^3(1.57) = .745\ldots$$
$$S^4(1.57) = .678\ldots$$

$$\vdots$$

$$S^{17}(1.57) = .385\ldots$$
$$S^{18}(1.57) = .375\ldots$$
$$S^{19}(1.57) = .366\ldots$$
$$S^{20}(1.57) = .358\ldots$$

Slowly, ever so slowly, successive iterates of $\sin x$ tend to 0:

$$S^{73}(1.57) = .185\ldots$$
$$S^{74}(1.57) = .196\ldots$$
$$S^{75}(1.57) = .194\ldots$$

$$\vdots$$

$$S^{148}(1.57) = .1402\ldots$$
$$S^{149}(1.57) = .1398\ldots$$
$$S^{150}(1.57) = .1393\ldots$$

$$\vdots$$

$$S^{298}(1.57) = .099527\ldots$$
$$S^{299}(1.57) = .099362\ldots$$
$$S^{300}(1.57) = .099199\ldots$$

So the orbit of $x = 1.57$ is the sequence of numbers $1.57, .999\ldots, .841\ldots$, $.745\ldots$, and we have

$$S^n(1.57) \to 0 \quad \text{as} \quad n \to \infty$$

That is, the orbit of 1.57 tends to 0.

Remark. To compute the orbit of 1.57, we have assumed that the sine function accepts numbers in radians rather than in degrees as input. On your calculator, there should be a button which allows you to select either radian mode or degree mode. In this book we will always assume that inputs to the sine and cosine functions are given in radians. If you use degrees, you will often get different answers from ours. Also, some calculators are in degree mode when they are turned on; others are in radian mode. You should always check which mode your calculator is using.

Exercise 1.3 Verify that the orbit of x under the sine function tends to 0 no matter what x is initially input. That is, check that

$$S^n(x) \to 0 \quad \text{as} \quad n \to \infty$$

for many different x-values.

Exercise 1.4 List the first 15 points on the orbit of 0 for each of the following functions. Can you predict what happens for all subsequent iterations?
  a. $F(x) = x + 1$
  b. $F(x) = 2(x + 1)$
  c. $F(x) = \frac{1}{2}(x + 1)$
  d. $F(x) = x^2 - 1$
  e. $F(x) = x^2 - 2$

We call the process of understanding all of the orbits of a given dynamical system *orbit analysis*. Let's use orbit analysis to understand all of the orbits of another function, $F(x) = 2x$. Clearly, $F(0) = 0$, so $F^2(0) = 0, F^3(0) = 0,\ldots$. Each point on the orbit of 0 is 0. Using the terminology introduced earlier, 0 is a fixed point; it never moves under iteration. On the other hand, if $x > 0$, then we have

$$F^n(x) \to \infty \quad \text{as} \quad n \to \infty$$

Indeed, we have

$$F(x) = 2x$$
$$F^2(x) = 4x$$
$$F^3(x) = 8x$$
$$\vdots$$
$$F^n(x) = 2^n x$$

and these numbers clearly grow larger and larger as n increases. If x is a negative, then these numbers become large negative numbers. Therefore, orbit analysis of $F(x) = 2x$ yields

1. $F^n(x) \to +\infty$ if $x > 0$
2. $F^n(x) \to -\infty$ if $x < 0$
3. $F^n(x) = 0$ if $x = 0$

Thus we know what happens to the orbits of all points under iteration of $F(x) = 2x$.

Exercise 1.5 Use orbit analysis to understand the behavior of all orbits of each of the following functions:

a. $F(x) = \frac{1}{2}x$
b. $F(x) = -3x$
c. $F(x) = -x$
d. $F(x) = x^3$
e. $F(x) = -x^2$

This brings us to the basic question in dynamical systems: Can we predict the fate of all orbits under iteration? Can we predict ahead of time what will happen when we iterate a function?

For all of the systems we have discussed using the calculator, namely, $\sqrt{x}$, x^2, and $\sin x$, the answer has been yes. Here is one last example where the fate of orbits can be decided, but the result is not so easy to predict ahead of time. What happens when the cosine function is iterated? Let's see. Let $C(x) = \cos x$ and choose any input, say $x = 18.84$ in radians. Then we find

$$C(18.84) = .999\ldots^*$$
$$C^2(18.84) = .540\ldots$$
$$C^3(18.84) = .857\ldots$$
$$C^4(18.84) = .654\ldots$$
$$C^5(18.84) = .793\ldots$$

$$\vdots$$

$$C^{99}(18.84) = .739085\ldots$$
$$C^{100}(18.84) = .739085\ldots$$
$$C^{101}(18.84) = .739085\ldots$$

So the orbit of 18.84 is a sequence of numbers that tends to .739085...

Experiment 1.6 Use a calculator to compute the orbits of various points under iteration of the cosine function.

Outcome. Again, no matter which x is initially input into the calculator, the result is the same: all orbits tend to .739085... Remember, you get a different answer if you use degrees. We'll see later why we always get this strange result.

1.5   Fixed and Periodic Points

At this point you might begin to suspect that the only types of orbits in a dynamical system are the fixed points and the orbits which tend to them (or to infinity). However, this is far from the truth: There are many, many different types of orbits in a typical dynamical system. Undoubtedly the most important type of orbit is a fixed point. Recall that a point x_0 is called a *fixed point* for F if $F(x_0) = x_0$. Note that fixed points never move under iteration: Since $F(x_0) = x_0$, it follows that $F(F(x_0)) = F(x_0) = x_0$ and, in general, $F^n(x_0) = x_0$. For example, as we saw in the previous sections, both 0 and 1 are fixed points for $S(x) = \sqrt{x}$ and $T(x) = x^2$. Similarly, 0 is a fixed point for $S(x) = \sin x$, while the point we found experimentally above, .739085..., is fixed by cosine.

Another type of orbit is the *periodic orbit*, or *cycle*. An orbit is periodic if it eventually returns to where it began. That is, the orbit of x_0 is periodic if there is an integer N such that $F^N(x_0) = x_0$. The point x_0 is called a periodic point of period N. The least such positive integer N is called the *prime period* of the orbit.

As an example, consider the reciprocal function $R(x) = 1/x$. We may input any $x \neq 0$ into this function. Note that $x = 1$ and $x = -1$ are both fixed points for R, since $R(1) = 1$ and $R(-1) = -1$. However, any other initial x generates a cycle of period 2. Indeed, if $x \neq \pm 1$ and $x \neq 0$, we have

$$R(x) = \frac{1}{x} \neq x$$

and

$$R^2(x) = R(1/x) = \frac{1}{\frac{1}{x}} = x$$

Hence x and $R(x)$ lie on the same periodic orbit for R.

A similar pattern occurs for the function $N(x) = -x$. Clearly, 0 is a fixed point for N, but all other points lie on a cycle of period two.

Exercise 1.7 Consider $F(x) = -x^3$. Can you find cycles of period 2 for F? Consider also $G(x) = (x + 1)(-\frac{3}{2}x + 1)$. Show that 0 is a periodic point for G. What is its prime period? What is the orbit of 0?

Notice several important properties of periodic orbits. Suppose x_0 lies on an orbit that is a cycle of period 4. We may write

$$x_1 = F(x_0)$$
$$x_2 = F(x_1) = F^2(x_0)$$
$$x_3 = F(x_2) = F^3(x_0)$$
$$x_0 = F(x_3) = F^4(x_0)$$

since x_0 has period 4. The orbit of x_0 therefore repeats cyclically

$$x_0, x_1, x_2, x_3, x_0, x_1, x_2, x_3, x_0, \ldots$$

Also, what about the orbits of x_1, x_2, and x_3? They too are cycles, since we know, for example, that

$$x_2 = F(x_1)$$
$$x_3 = F^2(x_1)$$
$$x_0 = F^3(x_1)$$
$$x_1 = F^4(x_1)$$

That is, the orbit of x_1 also repeats over and over again:

$$x_1, x_2, x_3, x_0, x_1, x_2, x_3, x_0, x_1, \ldots$$

As a consequence, each point on a cycle of period 4 for F will be fixed by F^4 as well as by F^8, F^{12}, and, in general, by F^{4n} for any integer n. All these multiples of 4 are called periods of the cycle. However, we reserve the term prime period for 4, the least period of the cycle.

The importance of periodic orbits or cycles stems from the fact that they represent cyclic or repeating phenomena in nature, such as the seasonal fluctuations of the populations of certain animal or insect populations. Incidentally, there is nothing special about our choice of a cycle of period 4; a cycle of any period has similar properties.

One final important type of orbit is the *eventually fixed* or *eventually periodic* orbit. These are points whose orbit is not fixed or periodic but for which some later point on the orbit is fixed or periodic. For example, the point $x = -1$ is eventually fixed for $T(x) = x^2$. This is true since $T(-1) \neq -1$ (so -1 is not a fixed point), but $T(-1) = 1$, and 1 is a fixed point. Similarly, the points $x = \pi$, 2π, $3\pi \ldots$ are all eventually fixed for $S(x) = \sin x$, since 0 is fixed by S, and

$$0 = S(\pi) = S(2\pi) = S(3\pi) = \cdots$$

Finally, the point 1 is eventually periodic for $F(x) = x^4 - 1$ because $F(1) = 0$ and 0 lies on a cycle of period 2.

Exercise 1.8 For each of the following functions, decide whether 0 is a fixed point, lies on a cycle, or is eventually periodic.

    a. $F(x) = |x|$
    b. $F(x) = 1 - x^2$
    c. $F(x) = (x - 1)^2$
    d. $F(x) = -\frac{1}{2}(x - 2)(3x + 1)$
    e. $F(x) = x^2 - 2$
    f. $F(x) = \pi \cos x$
    g. $F(x) = x^2 - 2x - 1$

1.6   An Application from Ecology

It is not our aim in this book to present applications of dynamical systems theory. Rather, we will content ourselves with presenting the mathematics, which, as we will see, becomes quite interesting in its own right. But, for the reader who insists on knowing why anyone would ever dream of iterating a simple function, we provide here one elementary application that arises in the study of population dynamics. We will see that this simple problem motivates much of the mathematics that comes later.

For a population biologist or ecologist who studies the growth or decline of the populations of different species of birds or fish or animals, an important problem is the construction of good mathematical models that will allow him or her to predict accurately the population in future generations or years. Will the population become extinct? Will there be a population explosion? Will the population fluctuate cyclically or behave totally unpredictably?

To answer these questions, the ecologist resorts to one of many mathematical models that are designed to aid in predicting population growth or decline. Most often, these models yield dynamical systems of one type or another.

One of the simplest dynamical systems that arises in this way is the logistic equation. This idealized model may be used to describe the behavior of the population of a single species that lives, reproduces, and dies in a controlled environment such as a laboratory without any unexpected external influences. This species may be a colony of boll weevils or ants or Alaskan snow frogs — the exact nature of the beast need not concern us here. Let's suppose that the ecologist can count accurately the population of the species during each generation. Then the question becomes what happens to the population as the generations go on? Can the ecologist use his complete knowledge of the present population to predict in advance the population many generations from now?

To keep the numbers manageable, let's keep track of only the percentage of some limiting population that is alive during each generation. In any environment, there is a maximum number of the species that can be present at any one time. This limiting population is governed, for example, by the physical size of the laboratory or colony in which the species is confined. After all, there can be no more of the species alive than can physically fit into the laboratory. We will take this number to be our maximum or limiting population. Let's call this number L.

The ecologist then denotes by P_n the percentage of this limiting population that is alive at generation n. So, for example, if $P_n = .5$, then the exact population is $L/2$ at generation n. Clearly, $0 \leq P_n \leq 1$. The logistic equation allows the ecologist to compute the population at generation $n + 1$ from a knowledge of the population in the preceding generation, P_n. The equation is

$$P_{n+1} = cP_n(1 - P_n)$$

Here c is an ecological constant that depends upon such things as the amount of food available or the temperature of the lab. The constant c is determined once and for all by the ecologist. For reasons we will discuss later, c is usually chosen between 0 and 4. Using this equation, and given any initial population P_0, the ecologist may then predict the species' population at any succeeding generation. For example, if $c = 2.4$ and the starting population is $P_0 = .05$, then

$$P_1 = 2.4(.05)(1 - .05) = .114$$
$$P_2 = 2.4(.114)(1 - .114) = .2424096$$

and so forth. That is, given any P_0, we may compute

$$P_1 = cP_0(1 - P_0)$$

The result of this computation allows us to compute P_2, P_3, and so on:

$$P_2 = cP_1(1 - P_1)$$
$$P_3 = cP_2(1 - P_2)$$
$$\vdots$$
$$P_n = cP_{n-1}(1 - P_{n-1})$$

Note that this process is simply the process of iterating the quadratic function $F(x) = cx(1 - x)$. Indeed, given P_0, we have

$$P_1 = F(P_0)$$
$$P_2 = F(P_1) = F^2(P_0)$$
$$\vdots$$
$$P_n = F(P_{n-1}) = F^n(P_0)$$

Therefore, the question asked by the ecologist is exactly the same as our original question: Predicting the fate of the population is the same as prediction of the fate of orbits generated by $F(x) = cx(1 - x)$. These functions are called the logistic functions.

A couple of remarks are in order. First, the logistic equation, as simple as it is, is a reasonable first approximation to a mathematical model for population growth. For example, if there is no species present $(P_0 = 0)$ or if the lab is completely full $(P_0 = 1)$, then there is no species present during the ensuing generations. This is true, since we have $F(0) = F(1) = 0$. Furthermore, if P_0 is small, the population tends to increase, whereas if P_0 is large, the population tends to decrease, as we would expect. On the other hand, it is highly unlikely that all the complexities of life and death can be mirrored in as simple an equation as the logistic equation, so this equation has to be taken with a grain of salt. Ecologists use much more sophisticated mathematical tools to make actual predictions. Nevertheless, as we shall see, this simple model leads to all kinds of unexpectedly complicated behavior. Iteration of the logistic function turns out to be one of the richest and most interesting examples of a dynamical system imaginable, and we return to it often in succeeding chapters.

Further Exercises and Experiments

1. Use a scientific calculator to find the first five points on the orbit of x_0 for each of the following functions.
   a. $F(x) = x^2 + 1$, $x_0 = 0, 1, 2$
   b. $G(x) = 3x - 1$, $x_0 = 1, 0.5$
   c. $H(x) = \frac{x}{3} - 1$, $x_0 = 3, 2, -3$

2. Use a scientific calculator to iterate each of the following functions. Some of these iterations demand more than one keystroke to compute each point on the orbit. Can you predict the fate of all orbits?

  a. $E(x) = \exp x$
  b. $S(x) = 1.5 \sin x$
  c. $R(x) = 1/x$
  d. $T(x) = 1/x^2$
  e. $U(x) = 1/\sqrt{x}$
  f. $A(x) = \arctan x$ (inverse tangent of x)
  g. $Q(x) = x^2 - 1$

3. Use a scientific calculator to find as many fixed points and periodic cycles as possible for each of the following functions.
  a. $Q(x) = x^2 - 1$
  b. $A(x) = 2 \arctan x$
  c. $S(x) = -1.5 \sin x$
  d. $L(x) = 2 - .5x$
  e. $C(x) = \cos x$

4. Perform orbit analysis on each of the following functions.
  a. $L(x) = -\frac{1}{2}x$
  b. $M(x) = -3x$
  c. $N(x) = x + 4$
  d. $P(x) = 4/x$
  e. $Q(x) = 2x - 1$
  f. $R(x) = x^4$
  g. $S(x) = -x^4$
  h. $T(x) = -x^3$

Chapter 2

Iteration Using the Computer

In this chapter we turn to the computer and computer graphics to help us understand iteration of functions. As is apparent, most functions are not represented by keys on a calculator and so iteration with a calculator is not often feasible. For example, to iterate a simple function such as the logistic function $2x(1 - x)$, we would have to "remember" the value of x (or store it in memory) while $1 - x$ is being computed in order to compute the product $2x(1 - x)$ subsequently. Such tasks are much easier to handle with a computer or programmable calculator. One day, someone might mass produce a calculator with a $2x(1-x)$ button for some crazy reason, but until that time, we must be content with programming the iterations ourselves. Besides, there are many, many more interesting functions to iterate than can ever be accommodated by the limited number of buttons on a calculator!

2.1   The Program ITERATE1

BASIC programs to iterate simple functions are not very difficult to write. They involve only one simple loop, which iterates the given function on a given initial input. For example, Figure 2.1 displays a BASIC program called ITERATE1 to iterate the function $2x(1-x)$ 25 times on a given initial input x_0. The program prompts the user to enter a value for x_0 and then prints out the next 25 points on the orbit of x_0. Incidentally, note that this function is one of the logistic functions we described in the last chapter, where $c = 2$.

```
REM  program ITERATE1
INPUT  "x0"; x0
FOR  i = 1  TO  25
     x1 = 2*x0*(1-x0)
     PRINT  i, x1
     x0 = x1
NEXT  i
END
```

Figure 2.1 The program ITERATE1.

Note how simple this program is. We use only the INPUT, FOR-NEXT, and PRINT statements from BASIC. When the program is run, the INPUT statement prompts us to type in an initial seed x_0. The computer then computes and displays the next 25 points on the orbit of x_0. This is accomplished in the FOR-NEXT loop of the program. In this loop, we first compute the value of the function

$$x_1 = 2x_0(1 - x_0)$$

Then we print both i, the iteration count, and x_1. Finally, we replace x_1 by x_0 and return to the beginning of the loop. Using this program, we can compute the successive iterates of this logistic function easily; computers are ideally suited for performing the routine task of iterating a function over and over again. They also save your index finger from developing blisters caused by computing long orbits on a calculator.

Remark. In ITERATE1 and in later programs, we will often use X0 as one of our variables. This is "X-zero," not "X-oh," as it is sometimes erroneously read. Also, the program could be simplified a bit by using the replacement statement

$$X0 = 2 * X0 * (1 - X0)$$

This statement combines the pair of statements

$$X1 = 2 * X0 * (1 - X0)$$
$$X0 = X1$$

into one.

Experiment 2.1 Use ITERATE1 to compute the iterates of various starting x_0-values. What happens? Can you use the output of this program to decide what happens to the orbit of any given starting x_0-value?

Outcome. The points $x_0 = 0$ and $x_0 = .5$ are fixed points. The point $x_0 = 1$ is taken onto 0 after one iteration, so 1 is eventually fixed. All other points satisfying $0 < x_0 < 1$ tend to .5 under iteration. When $x_0 < 0$ or $x_0 > 1$, the situation is quite different: all orbits tend to $-\infty$, as the reader will quickly notice because of the overflow messages that fill up the screen. We will see why all of this occurs in the next chapter when we discuss graphical analysis.

Project 2.2 Modify ITERATE1 so that it prints the first 25 points on the orbit of x_0, this time including x_0. Also, modify ITERATE1 so that it accepts from the user a new maximum number of iterations, MAXITER, rather than 25. Use another INPUT statement.

In this book, we present all programs using BASIC programming language. We urge readers who know other languages such as PASCAL or C to rewrite the programs in these languages. In general, programs written in these languages will run significantly faster. While speed is not an issue for a simple program like ITERATE1, it will become much more important later when we write programs that may take hours or even days to run!

Note that it is very simple to modify ITERATE1 so that the program computes the orbits generated by other functions. For example, if we change the fourth line of the program to read

$$X1 = 3 * X0 + SQR(X0)$$

or

$$X1 = SIN(X0) + COS(X0)$$

we will compute the orbits of the functions $F(x) = 3x + \sqrt{x}$ or $G(x) = \sin x + \cos x$, respectively. It is also easy to modify the number of points on each orbit that are computed by changing the 25 in the third line to any desired value.

Project 2.3 Modify ITERATE1 so that the program computes the iterates of each of the functions discussed in Chapter 1:

   a. $\sqrt{x}$
   b. x^2
   c. $\sin x$

  d. $\cos x$

Then use the computer to perform orbit analysis for each of these functions.

Clearly, ITERATE1 is a versatile program: A simple change in the function definition allows us to experiment with completely different dynamical systems.

Experiment 2.4 Use ITERATE1 to help you perform orbit analysis on the following.
  a. $F(x) = (\pi/2)\sin x$
  b. $F(x) = \exp x$
  c. $F(x) = -\cos x$
  d. $F(x) = \sin(x-1)$

2.2   The Logistic Function

We now begin one of the most fascinating subjects in all of dynamics: We use ITERATE1 to investigate the dynamics of the logistic function described in Chapter 1. Recall that the logistic function is the quadratic function given by

$$F(x) = cx(1-x) = c(x-x^2)$$

Here, c is a constant, which is usually chosen between 0 and 4. In ecology, the initial seed x_0 for the iteration is a percentage of some limiting population, so x_0 is usually chosen between 0 and 1.

We urge you to undertake the following project and experiment because much of our subsequent work and many of the major themes of this book will make use of this project.

Project 2.5 Modify ITERATE1 to compute the iterates of $cx(1-x)$ where c is a positive constant. Your new program should accept the following as input.
  a. Any desired value of the constant c (use an INPUT statement).
  b. Any initial point x_0 (use another INPUT statement).
  c. Any desired number of iterations, MAXITER (use a third INPUT statement).

When you input the values of MAXITER and c, it is best to keep MAXITER small, say between 100 and 200, and to keep c between 0 and 5.

Experiment 2.6 Now use this revised version of ITERATE1 to catalogue as many different dynamical phenomena as you can for the various logistic functions. Specifically, use this program to compute the iterates of any x_0 with $0 < x_0 < 1$. Record what you find. Try the following c-values, at the very least. If time permits, choose other intermediate c-values as well. What do you observe? You should record the results of your observations for comparison with the results of later experiments.

    a. $c = .5$
    b. $c = .8$
    c. $c = 1$
    d. $c = 1.5$
    e. $c = 2$
    f. $c = 3$
    g. $c = 3.2$
    h. $c = 3.5$
    i. $c = 3.55$
    j. $c = 3.83$
    k. $c = 4$
    l. $c = 5$

You should try other values of c to help find a more complete picture of the dynamics.

Outcome. The outcomes of these experiments vary with c. We have listed the first 38 iterations of 0.5 for three c-values in Table 2.2. For a typical value of x_0 with $0 < x_0 < 1$, you should see the following behavior.

    a. For $c = .5$, all orbits tend to 0.
    b. For $c = .8$, all orbits tend to 0.
    c. For $c = 1$, all orbits tend to 0, but very slowly. You will have to increase MAXITER substantially to see this.
    d. For $c = 1.5$, all orbits tend to $\frac{1}{3}$.
    e. For $c = 2$, all orbits tend to $\frac{1}{2}$. We observed this in our first experiment in the last section.
    f. For $c = 3$, all orbits tend to $\frac{2}{3}$, but very slowly. Moreover, these orbits oscillate from one side of $\frac{2}{3}$ to the other as they approach $\frac{2}{3}$.
    g. For $c = 3.2$, all orbits approach the period 2 cycle .5130456... and .799456....
    h. For $c = 3.5$, all orbits approach a period 4 cycle given by .38282..., .82694..., .50088..., and .87500....

i. For $c = 3.55$, all orbits approach a period 8 cycle.

j. For $c = 3.83$, all orbits approach a period 3 cycle.

k. For $c = 4$, there is no pattern whatsoever for a given x_0. Some initial x_0's have very simple orbits, such as $x_0 = \frac{1}{2}$. The orbit of $x_0 = \frac{1}{2}$ is

$$\frac{1}{2} \to 1 \to 0 \to 0 \to 0 \to \cdots$$

but the orbit of .501 is quite different: It never becomes fixed. Table 2.3 displays a table of outcomes of this experiment for various x_0 values when $c = 4$. Each point on the orbit is listed up to only three decimal places. Note that there is very little pattern to be discerned. We should remark that the table of values that your computer generates may not be the same as ours, at least after the first few iterations. This is a consequence of the different ways that different computers round off numbers as well as the "chaotic" nature of this iteration. We return to this point later.

l. For $c = 5$, all orbits apparently tend to $-\infty$. We say "apparently" for, as we shall see, there are actually many orbits that do not behave in this manner, although they are difficult to see on the computer.

As a consequence of this experiment, we see that even a very simple dynamical system — as simple as a quadratic function — can exhibit fairly complicated behavior. Moreover, the behavior changes as the constant c changes. This constant is called a *parameter*. For each different parameter value, we get a new function to iterate. Thus our experiment asked for the behavior of a family of dynamical systems for different parameter values.

Remark. If you try other parameter values in the preceding experiment, you will undoubtedly find many outcomes that are not listed. The more you experiment with the logistic functions, the more complex the dynamics will appear to become. It might even seem to be a hopeless task to understand everything that occurs! But don't be discouraged: After years of work, mathematicians have succeeded in understanding the major features of the dynamics of these functions only in the late 1970s. Even with the world's fastest computers at hand, understanding all that happens in the logistic family is a formidable task!

Exercise 2.7 Use ITERATE1 to find (at least approximately) the values of the parameter c for which all orbits of the function $F(x) = cx(1 - x)$

Iterate	$c = 1.5$	$c = 3.2$	$c = 3.5$
1	.375	0.8	.875
2	.3515625	.512	.3828125
3	.341949462	.7995392	.826934814
4	.337530041	.512884056	.500897694
5	.335405268	.799468803	.874997179
6	.334362861	.513018994	.382819903
7	.333846507	.799457618	.826940887
8	.333589525	.513040431	.500883795
9	.33346133	.79945583	.874997266
10	.333397307	.513043857	.382819676
11	.333365314	.799455544	.826940701
12	.333349322	.513044405	.500884222
13	.333341327	.799455499	.874997263
14	.33333733	.513044492	.382819683
15	.333335331	.799455491	.826940706
16	.333334332	.513044506	.500884209
17	.333333832	.79945549	.874997263
18	.333333583	.513044509	.382819683
19	.333333458	.79945549	.826940706
20	.333333395	.513044509	.50088421
21	.333333364	.79945549	.874997263
22	.333333348	.513044509	.382819683
23	.333333341	.79945549	.826940706
24	.333333337	.513044509	.50088421
25	.333333335	.79945549	.874997263
26	.333333334	.513044509	.382819683
27	.333333333	.79945549	.826940706
28	.333333333	.513044509	.50088421
29	.333333333	.79945549	.874997263
30	.333333333	.513044509	.382819683
31	.333333333	.79945549	.826940706
32	.333333333	.513044509	.50088421
33	.333333333	.79945549	.874997263
34	.333333333	.513044509	.382819683
35	.333333333	.79945549	.826940706
36	.333333333	.513044509	.50088421
37	.333333333	.79945549	.874997263
38	.333333333	.513044509	.382819683

Table 2.2 The orbit of .5 for various c-values. This orbit is attracted to a fixed point when $c = 1.5$, to a cycle of period 2 when $c = 3.2$, and to a cycle of period 4 when $c = 3.5$.

$x_0 =$	.1	.25	.3	.5	.51	.749	.8
1	.36	.75	.84	1	.999	.752	.640
2	.922	.75	.538	0	.001	.746	.922
3	.289	.75	.994	0	.006	.758	.289
4	.822	.75	.022	0	.025	.734	.822
5	.585	.75	.088	0	.099	.781	.585
6	.971	.75	.321	0	.356	.684	.971
7	.113	.75	.871	0	.917	.865	.113
8	.402	.75	.448	0	.301	.466	.402
9	.961	.75	.989	0	.842	.995	.962
10	.148	.75	.043	0	.530	.018	.148
11	.504	.75	.166	0	.996	.071	.504
12	1.000	.75	.554	0	.014	.262	1.000
13	.000	.75	.988	0	.058	.774	.000
14	.001	.75	.045	0	.219	.699	.001
15	.004	.75	.174	0	.686	.841	.004
16	.015	.75	.575	0	.861	.534	.015
17	.060	.75	.977	0	.477	.995	.059
18	.227	.75	.089	0	.998	.017	.222
19	.703	.75	.321	0	.007	.073	.690
20	.836	.75	.873	0	.031	.272	.856
21	.550	.75	.443	0	.120	.792	.494
22	.990	.75	.982	0	.423	.659	.999
23	.039	.75	.050	0	.976	.899	.000
24	.150	.75	.192	0	.091	.364	.002
25	.509	.75	.619	0	.333	.926	.008
26	.999	.75	.943	0	.889	.273	.033
27	.001	.75	.215	0	.393	.793	.129
28	.005	.75	.676	0	.954	.656	.450
29	.021	.75	.877	0	.174	.902	.990
30	.084	.75	.433	0	.575	.353	.040
31	.308	.75	.982	0	.977	.913	.154
32	.853	.75	.007	0	.082	.316	.521
33	.501	.75	.262	0	.321	.865	.998
34	.999	.75	.774	0	.872	.468	.008
35	.000	.75	.699	0	.443	.996	.031
36	.000	.75	.840	0	.987	.002	.118
37	.000	.75	.535	0	.049	.007	.418
38	.002	.75	.995	0	.189	.244	.973
39	.007	.75	.019	0	.614	.739	.105

Table 2.3 Various orbits of $4x(1 - x)$.

   a. Stop tending to 0 and begin tending to a different fixed point.

   b. Stop tending to a fixed point and start tending to a period 2 cycle.

   c. Stop tending to a period 2 cycle and start tending to a period 4 cycle.
Use only x_0-values between 0 and 1.

To be honest, it is difficult to find the exact c-values numerically where these changes occur. The study of where dynamical behavior changes as a parameter is varied is called *bifurcation theory*, a subject to which we will return quite often later in this book.

It may appear that the large number of different behaviors that we observed in the previous experiment were due to the special quadratic nature of the logistic function. This, however, is not true. Many, many different functions exhibit the same dynamical patterns as parameters are varied. Try the following experiments and see to what extent your results agree with observations about the logistic function.

Experiment 2.8 Consider the family of functions $S(x) = d\sin(x)$, where the parameter $d > 0$. Modify your program so that orbits of this family are computed. For $0 < d < 4$, can you find dynamical behavior for this family that is similar to that of the logistic family? As in the previous experiment, can you find d-values for which 0 attracts all orbits, for which there is a period 2 cycle, for which there is a period 4 cycle, and so forth? It is best to restrict the x_0-values to $0 < x_0 < \pi$ instead of $0 < x_0 < 1$. Record the values of d for which specific behavior is observed; they will be different from those observed for the logistic function. However, the observed phenomena should occur in the same order relative to these parameters.

Experiment 2.9 Consider the family of functions $F(x) = d(x - x^3)$ with $0 < d < 2.6$. Can you find similar dynamical behavior as we observed before for this family? Use initial x_0-values that satisfy $-1 < x_0 < 0$. Again record your observations.

2.3   Computer Graphics

As should be obvious by now, the computer is an indispensable tool for studying dynamical systems. The speed with which the computer generates orbits cannot be beat! But, somehow, the list of numbers that the program ITERATE1 prints is, aesthetically speaking, less than appealing. This goes beyond mere aesthetics. If, for example, our dynamical system possesses a cycle of period 100, it will be very hard to uncover this information from a listing of the orbit of a given x_0 value. Therefore, it is useful to seek another means of displaying the orbits of a dynamical system. One of the most effective ways of achieving this is by means of computer graphics. Instead of listing all of the points or numbers in an orbit, we instead plot the points of the orbit as they are computed on the computer screen. Just as "one picture is worth a thousand words," one graphics image can contain the same amount of information as a list of 100,000 points on an orbit of a dynamical system. Using graphics, we will be able to read the behavior of orbits quite easily.

To accomplish this, we need to be able to transfer information concerning the orbit of a dynamical system to the graphics screen. The simplest way to do this is to plot each point on the orbit in succession on the screen. This is easier said than done on many computers, for it necessitates transferring the information about orbits from the real line (where the dynamics are occurring) to the screen. That means we must change our "coordinates" from the place where the dynamics are occurring to the screen.

To do this, we must understand how the particular computer screen we are using is configured. A computer screen consists of a rectangular array of dots or pixels that we may color at will. Sometimes this means we may choose one of many colors to light up the pixel; other times, for monochrome displays, we will be able to choose only two colors, white and black. In any case, we need to know how to *address*, or name, each pixel on the screen. In this section, we will assume for illustration that the screen is a square array of 300 by 300 pixels. Most screens are *not* of this size, so the programs below will need to be modified to fit the appropriate screen. Usually, this is no problem, and the modifications can be carried out easily.

Just as in the Cartesian plane, points on the screen are given two coordinates, a horizontal and vertical coordinate. We will denote the horizontal coordinate by m and the vertical coordinate by n, so each pixel has a name, or address, of the form (m, n). In many but not all computers, the vertical coordinates run in the reverse direction from those in the Cartesian plane. This is true for the Apple Macintosh and IBM personal computers, that is, the upper left hand corner is named $(0, 0)$, whereas the bottom right hand corner has coordinates $(300, 300)$. The top horizontal line contains pixels named $(m, 0)$, and the rightmost vertical line contains pixels named $(300, n)$. So the vertical coordinate increases as you move down the screen, while the horizontal coordinate increases as you move to the right. See Figure 2.4. You should check to see how your screen is configured.

Let's now write a program that successively displays the first 200 points

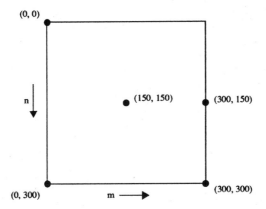

Figure 2.4 Screen coordinates.

of an orbit of the function $F(x) = 4x(1-x)$ along a horizontal line on the screen. As usual, we will choose an initial input x_0 satisfying $0 < x_0 < 1$. It turns out that each of the 200 points on the orbit of x_0 also lies in the interval between 0 and 1. For convenience, let's plot these points along the line on the screen whose vertical coordinate is 100. This means that points on the x-axis between 0 and 1 will correspond to pixels on the screen with label $(m, 100)$, where $0 \leq m \leq 300$.

Of course, m assumes only integer values, so not all the points on the x-axis will correspond exactly to a pixel on the line. For the moment, let's overlook this difficulty and ask how can we make points on the x-axis with $0 \leq x \leq 1$ correspond to points on the horizontal line $(m, 100)$ with $0 \leq m \leq 300$. The answer is given by the rule $m = 300x$. To any x with $0 \leq x \leq 1$, we associate the point $(m, 100)$ on the screen where $m = 300x$. Note that $x = 0$ corresponds to $(0, 100)$ under this association. This point is the leftmost point on our horizontal line. Similarly, $x = 1$ corresponds to $(300, 100)$, the rightmost point on the horizontal line. This procedure is called *changing coordinates*.

How did we find the rule $m = 300x$? Here is where some algebra helps out. We will always want our changes of coordinates to be as simple as possible. "Simple" here means that we want the formula relating m and x to be of the form

$$m = ax + b$$

where a and b are to be determined. But we know that $m = 0$ when $x = 0$,

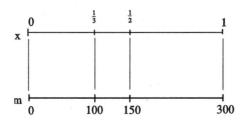

Figure 2.5 Changing coordinates.

so substituting this into the equation yields

$$0 = 0a + b$$

Therefore, $b = 0$. Similarly, we know that $m = 300$ when $x = 1$. Substituting again, we find

$$300 = 1a + 0$$

or $a = 300$. Thus $m = 300x$ is the correct change of coordinates.

Exercise 2.10 What change of coordinates would you use to convert the following intervals of x-values to screen coordinates $(m, 100)$ where $0 \leq m \leq 300$?

    a. $0 \leq x \leq 3$
    b. $0 \leq x \leq 2.5$
    c. $0 \leq x \leq \sqrt{5}$
    d. $1 \leq x \leq 2$
    e. $2 \leq x \leq 5$
    f. $-1 \leq x \leq 2$

There is an easy way to visualize this change of coordinates. In Figure 2.5 we have drawn two lines of equal length. One line — the x-axis — consists of points whose names run from 0 to 1. The other line — the m-axis — contains points whose names run from 0 to 300. The vertical arrows running from the x-axis to the m-axis give us the correspondence between points on these lines. Clearly, the point $x = \frac{1}{2}$ corresponds to the midpoint on the m-axis, namely, $m = 150$. Similarly, $x = \frac{1}{3}$ corresponds to $m = 100$ and $x = \frac{2}{3}$ corresponds to $m = 200$.

Figure 2.6 displays a program called ITERATE2. The aim of this program is to display the first 200 points on the orbit of x_0 under $F(x) =$

```
REM  program ITERATE2
INPUT  "x"; x0
CLS
FOR  i = 1 TO  200
     m = 300 * x0
     PSET (m, 100)
     x1 = 4 * x0 * (1 - x0)
     x0 = x1
NEXT i
END
```

Figure 2.6 The program ITERATE2.

$4x(1-x)$ where $0 < x_0 < 1$. In the program, the actual plotting is accomplished by the two commands

$$M = 300 * X0$$

$$PSET(M, 100)$$

The first line simply changes coordinates from x-coordinates to screen coordinates, while the second line tells the computer to light up the pixel $(m, 100)$. Recall that the number m is not usually an integer. The PSET command overcomes this difficulty by rounding m to the nearest integer before plotting the point. Thus we get only an approximation to the actual orbit when we use ITERATE2; we cannot in general distinguish two points whose distance apart is less than $1/300$.

The statement CLS in the program simply clears the screen before we begin plotting. This command and the PSET command may vary in other dialects of BASIC. You should check your version of BASIC for the appropriate commands.

The main computation in the program takes place within the FOR-NEXT loop via the statements

$$X1 = 4 * X0 * (1 - X0)$$
$$X0 = X1$$

These statements are familiar from our previous program, ITERATE1. Using ITERATE2, we may now "visualize" the orbits of $F(x) = 4x(1-x)$, as in the following experiment. You should compare the graphical data generated by this experiment to the table in Table 2.3 generated by ITERATE1.

Experiment 2.11 Use ITERATE2 to compute the orbits of various x_0 values for $F(x) = 4x(1 - x)$ when $0 < x < 1$. What do you see?

Outcome. Some orbits are quite simple — for example, the orbits of $\frac{1}{2}$, $\frac{3}{4}$, or $\frac{1}{4}$, for which only finitely many pixels are lit. Others seem to wander aimlessly about the interval, lighting up virtually every point in the interval. This is our second encounter with chaotic dynamics, a topic we investigate in detail later.

Exercise 2.12 Write a program similar to ITERATE2 that displays the first 300 points on an orbit of $S(x) = \pi \sin x$ for an initial x_0 in the interval $0 \le x_0 \le \pi$. Remember to use an appropriate change of coordinates.

ITERATE2 has several defects. One is the fact that all points on the orbit are displayed. Very often, we want to know only the eventual, or *asymptotic*, behavior of an orbit, that is, what happens to very high iterations.

Project 2.13 Modify ITERATE2 so that the first 100 points on the orbit are not displayed; only the last 100 points on the orbit are shown. Remember that you still must compute *all* 200 points on the orbit.

Experiment 2.14 Use this project to reinvestigate the dynamics of the logistic function $cx(1 - x)$ for various c-values. Do you "see" the same results that you found in Section 2.2 using ITERATE1?

A second defect of ITERATE2 is the fact that, once a pixel is lit, it is never extinguished. This makes it difficult to see the actual behavior of orbits; for example, cycles appear as a finite set of points on the screen. There are various ways to "erase" a pixel once it is lit. Here is one way to do this. Just as the command

$$\text{PSET}(M, 100)$$

colors a pixel black, the command

$$\text{PSET}(M, 100), 30$$

colors the same pixel white (or vice versa, depending upon your computer). Therefore, the statements

$$\text{PSET}(M, 100)$$
$$\text{FOR } J = 1 \text{ TO } 1000$$
$$\text{NEXT } J$$
$$\text{PSET}(M, 100), 30$$

when incorporated into ITERATE2, have the effect of lighting a pixel for a moment, and then extinguishing it. The empty loop

$$\text{FOR } J = 1 \text{ TO } 1000$$
$$\text{NEXT } J$$

causes the computer to pause as it counts to 1000, thereby allowing the pixel to remain lit for a moment or two, unless your computer is very slow, in which case the pixel may remain lit for longer. You can change the value 1000 to achieve any desired effect.

Project 2.15 Incorporate these changes into ITERATE2 so that points along the orbit are lit for a moment, and then extinguished. It might be helpful to plot a small box rather than a point at each iteration. It also is helpful to list below the axis some of the coordinates of points on the axis, so that the relative positions of points on the orbit may be estimated.

The next modification of ITERATE2 is important. We use this version of the program several times later in this book.

Project 2.16 For our later work, modify ITERATE2 so that it accepts as input an initial value of x_0 together with the endpoints of an interval $\ell \le x_0 \le r$ in which the orbit is plotted. Call this new program ITERATE3.

The only difficulty encountered in this modification involves the transformation from real coordinates to screen coordinates. Assuming that the values of ℓ and r are given and that we wish to convert to screen coordinates of the form $(M, 100)$ with $0 \le M \le 300$, the necessary transformation is

$$M = 300 * (x - \ell)/(r - \ell)$$

Note that when $x = \ell$, $M = 0$ and when $x = r$, $M = 300$, as required.

Exercise 2.17 What is the formula for the change of coordinates that takes the interval $2 \le x \le 4$ to screen coordinates $(M, 100)$ with $0 \le M \le 250$?

Further Exercises and Experiments

1. Modify ITERATE1 so that it computes the orbits of the following functions. Can you describe the fate of all orbits?

   a. $Q(x) = x^2 - 1$
   b. $Q(x) = x^2 - \frac{1}{4}$
   c. $T(x) = x^3$
   d. $T(x) = -x^3$
   e. $A(x) = \arctan x$ (ATN (x))
   f. $A(x) = 2 \arctan x$
   g. $E(x) = .3 \exp x$
   h. $E(x) = -3 \exp x$

2. Redo Experiment 2.6, this time using negative c-values for the parameter in the function $F(x) = cx(1 - x)$. Record your observations.

3. Use ITERATE1 to compute the orbit of the given x_0 for a variety of c-values in the specified range for each of the following functions. Do you see any similarities among these different functions?

   a. $Q(x) = x^2 + c$ for c decreasing from .5 to 0; $x_0 = 0$
   b. $E(x) = c \exp(x)$ for c decreasing from .4 to .3; $x_0 = 0$
   c. $H(x) = x - x^2 + c$ for c increasing from $-.2$ to $.2$; $x_0 = .5$

4. Use ITERATE1 to compute the orbit of the given x_0 for a variety of c-values in the specified range for each of the following functions. Do you see any similarities among these different functions?

   a. $Q(x) = x^2 + c$ for c decreasing from 0 to -1; $x_0 = 0$
   b. $A(x) = c \arctan x$ for c decreasing from $-.5$ to -1.5; $x_0 = 1$
   c. $E(x) = c \exp x$ for c decreasing from -2 to -3; $x_0 = 0$

5. Write a program similar to ITERATE2 that will display all points on an orbit of $Q(x) = x^2 - 2$ in the interval $-2 \le x \le 2$.

6. Use ITERATE3 to display the orbit of the given x_0 for the following functions over the specified intervals $\ell \le x_0 \le r$.

   a. $Q(x) = x^2 - 1.5$, $\ell = -2, r = 2$; $x_0 = 0$
   b. $S(x) = \pi \sin x$, $\ell = -\pi, r = 0$; $x_0 = 1$

Chapter 3

Graphical Analysis

The goal of this section is to combine a mathematical procedure with our previous experimental work to see why some of the dynamical behavior we discovered before occurs. We call this technique *graphical analysis*. Using only the graph of the function, we will be able to understand the behavior of the iterates of the function. We will also be able to follow orbits and perform orbit analysis geometrically, without resorting to computing the graphs of the higher iterates.

3.1   The Graph of a Function

One of the best ways to understand what a function does is to use its graph. The graph of a function is a concise picture of all of the values of the function presented in an easy-to-read way.

To construct the graph of a function, let's start with a given function F. In the Cartesian plane (the xy-plane), the graph of F is simply the set of all points of the form $(x, F(x))$. That is, for each allowable x, we record both the input x and the output $F(x)$ in the single point $(x, F(x))$ in the plane. For example, if $F(x) = x$, then the graph of F is simply the set of points of the form (x, x) in the plane. That is, the graph of F is the set of points whose x- and y-coordinates are the same. This, of course, is the straight line that makes a $45°$ angle with both the x- and y-axes. See Figure 3.1.

To construct the graph of a function, we need to plot the points $(x, F(x))$ for all allowable x-values. This is a time-consuming task when done by hand, but the computer makes the task much easier. Using the computer, we can

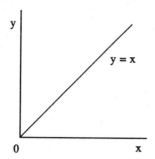

Figure 3.1 The graph of $F(x) = x$.

tabulate a list of inputs and their corresponding outputs, and then plot each point in the xy-plane. This gives a collection of dots in the plane which approximates the graph of the function.

As an alternative, you may use one of the many good software packages available to plot graphs on your computer screen quickly and easily. Using some of the graphing techniques that we discuss later, you can even construct your own graphing package, which is customized for your use. We will not take the time at this juncture to explain how to do this.

There are many graphs that you should be able to recognize and plot fairly quickly. Figure 3.2 gives several of them, including $F(x) = \sqrt{x}$, $F(x) = x^2$, $F(x) = 4x(1 - x)$, and $F(x) = x^2 - 2$.

The graph of F gives us lots of information about the first iterate of a function, but to understand the dynamics of F, we need to know about F^2, F^3, and so forth. Graphical analysis gives this information using only the graph of F.

3.2   Using Graphical Analysis

To explain graphical analysis, let us return to our old friend $S(x) = \sqrt{x}$. At first glance, it appears that we should need to know the graphs of not only $S(x)$, but also $S^2(x)$, $S^3(x)$, and so forth in order to understand the fate of all orbits. But this is not so: There is a simple geometric procedure for describing the behavior of orbits using only the graph of $S(x)$. Recall that the graph of $S(x)$ is given as in Figure 3.2a. To describe the orbit of a point x_0, we will first draw the diagonal line $y = x$, which makes a 45° angle with the x- and y-axes (see Figure 3.1.) The next point on the orbit of x_0 is the number $S(x_0)$. The graph of S allows us to read off this point, since

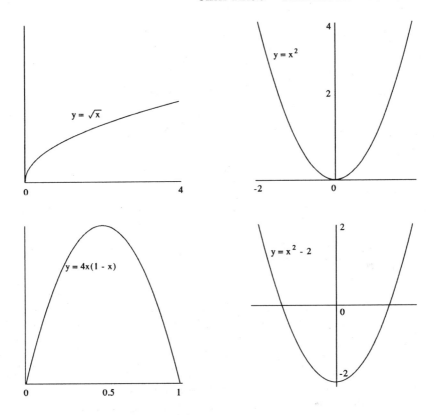

Figure 3.2 Some often-encountered graphs.

$(x_0, S(x_0))$ is the point on the graph directly over x_0. So if we draw a vertical line from the point (x_0, x_0) on the diagonal to the graph of S, the point of intersection has y-coordinate $S(x_0)$. Next we note that a horizontal line through this point intersects the y-axis at precisely $S(x_0)$. We would prefer to see this point on the same axis as x_0. To do this, we draw the horizontal line through $(x_0, S(x_0))$ to the diagonal instead. The point where this line meets the diagonal is precisely $(S(x_0), S(x_0))$. Therefore, the number $S(x_0)$ lies on the x-axis directly below this point. See Figure 3.3.

This gives us a procedure for finding the orbit of x_0. If we start on the diagonal at (x_0, x_0) and draw the vertical line to the graph at $(x_0, S(x_0))$, followed by the horizontal line back to the diagonal, we end up at $(S(x_0), S(x_0))$. Now do this again. A vertical line to the graph from $(S(x_0), S(x_0))$ hits the graph at $(S(x_0), S(S(x_0)))$. Then a horizontal line reaches the diagonal at $(S^2(x_0), S^2(x_0))$. Thus the orbit of x_0 is appearing along the diagonal. A

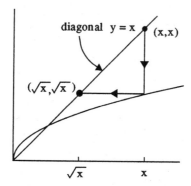

Figure 3.3 Graphical analysis of $S(x) = \sqrt{x}$.

picture explains this most easily. See Figure 3.4, which exhibits the orbits of two points under the square root function. Note that successive applications of graphical analysis show that these orbits tend to 1, a fact that we determined experimentally in Chapter 1. Also note that the vertical line from the diagonal to the graph may go up or down, depending upon the position of the graph relative to the diagonal. Similarly, the horizontal line from the graph back to the diagonal may go to the left or right. It is important to remember to draw the vertical line to the graph first, then the horizontal line back to the diagonal.

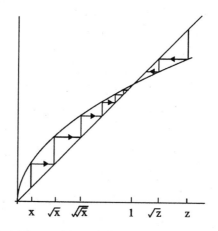

Figure 3.4 Two orbits of S given by graphical analysis.

Graphical analysis enables us to discover many of the elementary dynamical properties of a function. For example, Figure 3.5 illustrates graphical analysis applied to the squaring function $T(x) = x^2$. Note that the fixed points 0 and 1 correspond to points of intersection of the graph of T with the diagonal. All points x_0 with $|x_0| < 1$ have orbits that tend to 0, while points x_0 with $|x_0| > 1$ have orbits that tend to infinity.

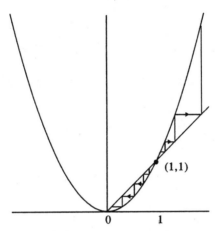

Figure 3.5 Graphical analysis of $T(x) = x^2$.

In Chapter 1, we saw that all orbits of the cosine function tended to the point .739085.... Graphical analysis explains why this happens. The graph of $C(x) = \cos x$ is shown in Figure 3.6. Note that there is a unique point of intersection of this graph with the diagonal $y = x$. This point has x-coordinate .739085..., although it is impossible to solve for this value exactly. Graphical analysis shows that all other orbits tend to this point. To be effective, graphical analysis necessitates an accurate graph. You should try your hand at a few simple examples.

Exercise 3.1 Using graphical analysis, describe the behavior of all orbits of each of the following functions.

   a. $F(x) = 2x$
   b. $F(x) = \frac{1}{2}x$
   c. $F(x) = 3 - 2x$
   d. $F(x) = x^3$

We can sometimes find periodic cycles using graphical analysis. For example, as we have seen in Chapter 1, the function $F(x) = -x^3$ has a cycle

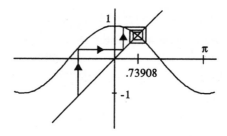

Figure 3.6 Graphical analysis of $C(x) = \cos x$.

of period 2 given by 1 and -1, since $F(1) = -1$ and $F(-1) = 1$. This cycle is represented by the box in Figure 3.7.

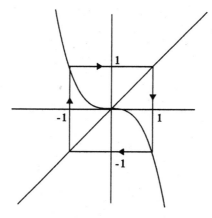

Figure 3.7 A cycle of period 2 for $F(x) = -x^3$.

Exercise 3.2 Use graphical analysis to show that 0 and -1 lie on a cycle of period 2 for the function $F(x) = x^2 - 1$.

Sometimes graphical analysis fails to predict the behavior of a function. For example, Figure 3.8 illustrates graphical analysis of $F(x) = 4x(1 - x)$. Note the complexity of the orbit depicted, a fact we observed experimentally in the last section. This example illustrates that long orbits are sometimes difficult to trace with this method.

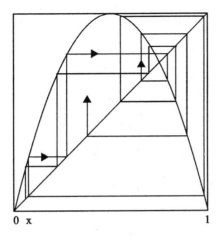

0 x   1

Figure 3.8 Graphical analysis of $F(x) = 4x(1 - x)$.

3.3   Attracting and Repelling Fixed Points

There is an important difference in the qualitative behavior near certain fixed points that may be readily explained using graphical analysis. Consider the two functions $H(x) = \frac{1}{2}x$ and $G(x) = 2x$. Both of these functions have 0 as a fixed point. But the dynamics near 0 are different in each case. For H, graphical analysis shows that all orbits tend to 0 under iteration: 0 attracts the orbits of all points. This, of course, can be readily verified using a calculator or one of our programs. We therefore call 0 an *attracting* fixed point for H. On the other hand, all nonzero orbits for G behave differently: They move away from 0. In this case 0 is a *repelling* fixed point. See Figure 3.9.

To be more precise, suppose a function F has a fixed point p. The point p is called *attracting* if there is an interval $a < x < b$ containing p in which all points have orbits that tend to p. That is, if x satisfies $a < x < b$, then $F^n(x) \to p$ as $n \to \infty$. In plain English, this means that points which are close enough to p (within the interval $a < x < b$) have orbits which tend to p.

For example, 0 is an attracting fixed point for $T(x) = x^2$, since all points in the interval $-1 < x < 1$ (excluding the end points) have orbits which tend to 0.

Attracting fixed points are often easy to detect experimentally using ITERATE2; all we need do is find a point that lies in the interval $a < x < b$

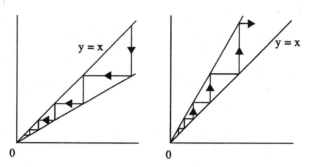

Figure 3.9 Graphical analysis of $H(x) = \frac{1}{2}x$
and $G(x) = 2x$.

and watch as its orbit tends to p. The only difficulty is locating the interval $a < x < b$ in the first place.

An important related concept is the *basin of attraction* of an attracting fixed point. The basin of attraction consists of all points whose orbits tend to a given attracting fixed point. For example, for $T(x) = x^2$, the basin of attraction of the attracting fixed point 0 is the interval $-1 < x < 1$. This same interval is the basin of attraction of 0 for the function $F(x) = x^3$, while all real numbers lie in the basin of attraction of 0 for $H(x) = \frac{1}{2}x$. Figure 3.10 shows how we may use graphical analysis to detect the basin of attraction of an attracting fixed point.

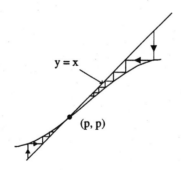

Figure 3.10 The point p is an attracting fixed point.

Experiment 3.3 Find the attracting fixed point of $P(x) = x^2 - \frac{1}{2}$. What is the basin of attraction of this fixed point? Use ITERATE2 to determine experimentally both the fixed point and the interval of attraction, and then calculate these values rigorously.

Outcome. Recall that a fixed point of F satisfies the equation $F(x) = x$. Hence the fixed points are given in this case by the equation

$$x^2 - \frac{1}{2} = x$$

or

$$x^2 - x - \frac{1}{2} = 0$$

This is a quadratic equation whose roots may be determined using the quadratic formula, which yields two roots,

$$x = \frac{1 - \sqrt{3}}{2} \quad \text{and} \quad x = \frac{1 + \sqrt{3}}{2}$$

Graphical analysis shows that $(1 - \sqrt{3})/2$ is attracting and the corresponding basin of attraction consists of all points x such that

$$\frac{-1 - \sqrt{3}}{2} < x < \frac{1 + \sqrt{3}}{2}$$

See Figure 3.11.

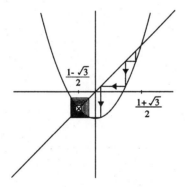

Figure 3.11 Graphical analysis of $P(x) = x^2 - \frac{1}{2}$.

Exercise 3.4 Determine both experimentally and via graphical analysis the attracting fixed points and their basins of attraction for each of the following functions:

    a. $T(x) = -x^3$

    b. $H(x) = 2x(1 - x)$

   c. $J(x) = 2.5x(1-x)$

   d. $K(x) = x^3 + \frac{1}{4}x$

   e. $S(x) = \frac{1}{2}\sin x$

The second kind of fixed point is a *repelling* fixed point. These fixed points have nearby orbits that behave in the exact opposite manner from attracting fixed points; instead of tending toward the fixed point under iteration, nearby points tend away from a repelling fixed point. To be precise, suppose F has a fixed point at p. This point is called a repelling fixed point if there is an interval $a < x < b$ containing p which has the property that all orbits (except p) leave the interval $a < x < b$ under iteration. That is, there is an interval $a < x < b$ with $a < p < b$ such that, if x satisfies $a < x < b$ and $x \neq p$, then $F^n(x)$ does not lie in $a < x < b$ for some n.

This definition is somewhat technical, but a few examples clarify what it says. For example, 1 is a repelling fixed point for $T(x) = x^2$. We may choose the interval $\frac{1}{2} < x < 2$ to see this. Clearly, all orbits except the fixed point eventually leave this interval under iteration. Indeed, if $\frac{1}{2} < x < 1$, then $T^n(x) \to 0$ as $n \to \infty$, and so the orbit of x eventually leaves the interval $\frac{1}{2} < x < 2$ and comes close to 0. Similarly, if $1 < x < 2$, then $T^n(x) \to \infty$ as $n \to \infty$ and, again, all of these orbits depart. There is nothing special about the interval $(\frac{1}{2}, 2)$; we could have chosen any interval (a, b) with $0 < a < 1$ and $b > 1$ to verify that 1 is a repelling fixed point.

Graphical analysis allows us to determine easily whether or not a given fixed point is repelling. Consider the function F whose graph is displayed in Figure 3.12. The fixed point p in this case is a repelling fixed point, since nearby orbits tend to move far away from p.

Exercise 3.5 Use graphical analysis to find all repelling fixed points for each of the following functions.

   a. $T(x) = x^3$

   b. $F(x) = 2x(1-x)$

   c. $V(x) = 1 - 2x$

It is more difficult to find repelling fixed points using the computer than it is to find attracting fixed points. For repelling fixed points, nearby orbits move *away* from the fixed point rather than toward it. Hence nearby orbits give us no clue that there is a fixed point in the vicinity. However, if we know the location of a fixed point at the outset, then we may use ITERATE1 or ITERATE2 to check whether it is repelling.

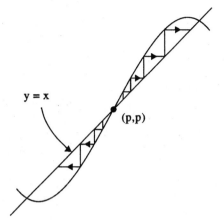

Figure 3.12 The point p is a repelling fixed point.

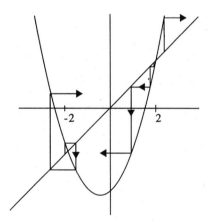

Figure 3.13 Graphical analysis of $F(x) = x^2 + x - 4$. Note that both fixed points are repelling.

For example, consider the function $F(x) = x^2 + x - 4$. This function has two fixed points, at $x = +2$ and $x = -2$. Are these fixed points attracting or repelling? Use the computer to check that, if you select an initial input sufficiently near ± 2, then the orbit moves away. Can you find an appropriate interval $a < -2 < b$ from which all orbits (except the fixed point) eventually leave? See Figure 3.13 for graphical analysis of this function.

3.4   Stable and Unstable Orbits

There is a fundamental difference between attracting and repelling periodic points: Attracting periodic points can be "seen" using the computer, whereas repelling periodic points cannot. We saw this when we tried to find the fixed points of the function $P(x) = x^2 - \frac{1}{2}$. There were two fixed points, one attracting, given by

$$x = \frac{1 - \sqrt{3}}{2}$$

and one repelling, given by

$$x = \frac{1 + \sqrt{3}}{2}$$

If we use ITERATE1 to try to find these fixed points, we see that the typical behavior of orbits is to tend to the attracting fixed point or to tend to ∞. We would have to be very lucky indeed to guess the location of the repelling fixed point.

This brings us to one of the principal themes of this book, the notions of stability and instability. An orbit of a dynamical system is called *stable* if it has the property that, if you change the initial input slightly, the resulting orbit behaves similarly. For example, all nonzero orbits of the square root function are stable because, as we have seen, they all tend to the attracting fixed point at 1. Similarly, for $C(x) = \cos x$, all orbits are again stable because they all tend to the attracting fixed point at .739085 ....

For the squaring function $T(x) = x^2$, if $|x| < 1$, then $T^n(x) \to 0$. So all orbits of points with $|x| < 1$ are stable. Similarly, if $|x| > 1$, then $|T^n(x)| \to \infty$, so again, all these orbits are stable. The only remaining orbits are those of 1 (a repelling fixed point) and -1 (an eventually fixed point). These orbits are *unstable* because nearby orbits have vastly different behaviors: they tend to 0 or to ∞ depending upon whether they are less than or greater than 1 in absolute value.

Notice that an attracting fixed or periodic point is always stable, whereas a repelling point is never stable (nearby initial conditions tend far away). It is also true that any point in the basin of attraction of an attracting fixed or periodic point is stable.

Exercise 3.6 For each of the following functions, determine which of the orbits are stable and which are unstable.

a. $F(x) = 2x + 1$
b. $G(x) = \frac{1}{3}x - 3$
c. $H(x) = x^3$
d. $J(x) = x^4$

Experiment 3.7 Use ITERATE1 to determine whether the orbit of the given x_0 is stable for the following functions:

a. $x_0 = 0$, $S(x) = 2\sin x$
b. $x_0 = 0$, $S(x) = -2\sin x$
c. $x_0 = \frac{1}{2}$, $Q(x) = 4x(1-x)$
d. $x_0 = 0$, $J(x) = x^2 - 3$
e. $x_0 = 1$, $Q(x) = 3x(1-x)$
f. $x_0 = 0$, $Q(x) = 2x(1-x)$

Outcome. All but one are unstable!

3.5   Attracting and Repelling Periodic Points

Like fixed points, periodic orbits may also be either attracting or repelling. For example, consider the function $F(x) = -x^3$. The points 1 and -1 lie on a cycle of period 2. This cycle is a repelling periodic orbit. You may see this in a number of ways. Using the computer and ITERATE1 or ITERATE2, you may easily check that if $|x| > 1$, then $|F^n(x)| \to \infty$. On the other hand, if $|x| < 1$, then $|F^n(x)| \to 0$. Hence any point near ± 1 has an orbit which tends far away.

Another way to see this is to use graphical analysis. This is shown in Figure 3.14. Note that all orbits of points with $|x| \neq 1$ behave in the manner just described.

Finally, we may also deduce that this cycle is repelling by working with $F^2(x)$. We have

$$F^2(x) = F(-x^3) = -(-x^3)^3 = x^9$$

The graph of $F^2(x)$ is shown in Figure 3.15. Note that both 1 and -1 are fixed points for F^2, and they are both clearly repelling. Since neither are fixed points for F, they must therefore lie on a cycle of period 2.

Exercise 3.8 The points $x = 0$ and $x = -1$ lie on a cycle of period 2 for $F(x) = x^2 - 1$. Is this cycle attracting or repelling? Use ITERATE1 or ITERATE2 to decide.

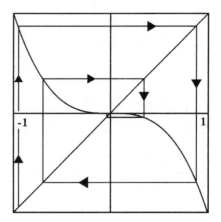

Figure 3.14 Graphical analysis of $F(x) = -x^3$.

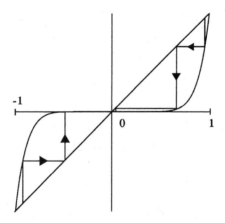

Figure 3.15 Graphical analysis of $F^2(x) = x^9$.

3.6   Higher Iterates

Another application of graphical analysis allows us to produce a rough sketch of the graph of F^n from a knowledge of the graph of F. Given x_0, graphical analysis gives us a quick method of finding (at least approximately) the point $(F^n(x_0), F^n(x_0))$ on the diagonal. So this gives us the y-coordinate of the point on the graph of F^n over x_0. If we apply graphical analysis to a sufficient number of points, we can generate a rough sketch of the graph of F^n.

Let's apply this technique to sketch the graph of some of the iterates of

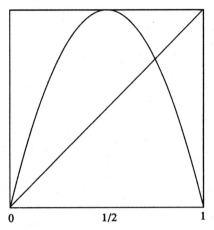

Figure 3.16. The graph of $F(x) = 4x(1 - x)$ for $0 \le x \le 1$.

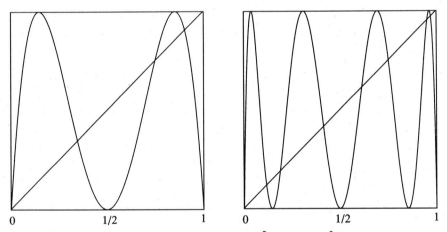

Figure 3.17. The graphs of $F^2(x)$ and $F^3(x)$.

$F(x) = 4x(1 - x)$ for $0 \le x \le 1$. The graph of F is shown in Figure 3.16. Note that $F(\frac{1}{2}) = 1$ and $F(0) = F(1) = 0$. The graph shows that F takes the intervals $0 \le x \le \frac{1}{2}$ and $\frac{1}{2} \le x \le 1$ onto the interval $0 \le x \le 1$. This means that there are points a_0 in $0 \le x \le \frac{1}{2}$ and b_0 in $\frac{1}{2} \le x \le 1$ that satisfy $F(a_0) = F(b_0) = \frac{1}{2}$. Hence $F^2(a_0) = F^2(b_0) = F(\frac{1}{2}) = 1$. On the other hand, F^2 takes each endpoint of $0 \le x \le \frac{1}{2}$ and $\frac{1}{2} \le x \le 1$ to 0. This means that the graph of F^2 has two humps in the interval $0 \le x \le 1$. See Figure 3.17.

Arguing similarly, there are points in each of the four intervals $0 < x <$

a_0, $a_0 < x < \frac{1}{2}$, $\frac{1}{2} < x < b_0$, and $b_0 < x < 1$ that are taken to $\frac{1}{2}$ by F^2. Hence F^3 takes each of these points to 1, whereas F takes the endpoints of the four intervals to 0. So the graph of F^3 has four humps — one in each of these intervals. See Figure 3.17.

Continuing in this fashion, we see that the graph of F^n has 2^{n-1} humps between 0 and 1. Each of these humps extends from $y = 0$ to $y = 1$, so, as a consequence, the graph of F^n crosses the diagonal at least 2^n times. This is an important fact; it means that F^n has at least 2^n fixed points. Not all of these points are periodic with prime period n; for example, both 0 and $\frac{3}{4}$ are fixed points for F (and so fixed also for F^n). But many of these points do have prime period n. This shows that a simple function like $F(x) = 4x(1-x)$ may have many, many periodic points with many different periods.

The importance of this observation is the following. Suppose we were to try to find out how many points are left fixed by F^4 by other means. Graphical analysis says that there are at least $2^4 = 16$ of them. (In fact, there are exactly 16 such points: 2 fixed points, a cycle of period 2, and 3 cycles of period 4.) One possible way to find these points would be to try to solve the equation

$$F^4(x) = x$$

algebraically. We challenge you to attempt this! This equation is a polynomial equation of degree 16. Write this out and check it for yourself. Two roots are easy to find, the fixed points $x = 0$ and $x = \frac{3}{4}$. But the other 14 roots are far from obvious. In fact, it can be proved that there is no general method to solve polynomial equations of degree greater than 4, so efforts in this direction are usually fruitless. Yet notice how quickly and effortlessly graphical analysis yields the existence of these cycles.

Lest you think that any polynomial equation of degree 16 always has 16 real roots, we remind you that $T(x) = x^2$ has only 2 fixed points and no other periodic points at all. The equation for period 4 cycles in this case is again a polynomial of degree 16, namely, $x^{16} = x$, but there are only two real solutions to this equation.

Another possible approach to finding periodic cycles for $F(x) = 4x(1-x)$ would be to use the computer and one of our earlier programs. At this point it is useful to recall some of the experiments from Chapter 2 in which we computed various orbits of $F(x) = 4x(1 - x)$. When performing these experiments, did you find any periodic points besides the two fixed points? Probably not. It is extremely difficult to find periodic points for this function using the computer because these orbits are unstable and are therefore usually invisible to the computer.

Further Exercises and Experiments

1. Use graphical analysis to describe the behavior of all orbits of the following functions:

   a. $F(x) = 2x + 1$
   b. $F(x) = -x + 2$
   c. $F(x) = x^4$
   d. $F(x) = x^2 + \frac{1}{4}$
   e. $F(x) = x + x^3$
   f. $F(x) = x - x^3$
   g. $F(x) = x + x^2$

2. Find all fixed points for each of the following functions and determine whether they are attracting or repelling.

   a. $F(x) = x^2 - \frac{1}{4}$
   b. $F(x) = x^2 - 1$
   c. $F(x) = x^2 - 2$
   d. $F(x) = 4x^3$
   e. $F(x) = 1/x^2$
   f. $F(x) = 1/\sqrt{x}$

   Find each fixed point explicitly, and then use ITERATE1 or ITERATE2 to determine whether this point is attracting or repelling. Use graphical analysis to explain these results.

3. There are certain fixed points that are neither attracting or repelling. These fixed points are called *neutral*. Find all neutral fixed points for each of the following functions:

   a. $F(x) = x$
   b. $F(x) = -x + 3$
   c. $F(x) = x + x^2$
   d. $F(x) = 1/x$
   e. $F(x) = x^2 + \frac{1}{4}$

4. Find all fixed points and periodic points of period 2 for $F(x) = x^2 - 1$. Determine whether these orbits are attracting or repelling using both the computer and graphical analysis.

5. Write a BASIC program that performs graphical analysis on a function. Your program should accept as input an x_0-value and then plot the vertical and horizontal lines of 20 successive iterates of x_0.

6. Use graphical analysis to decide for which values of c the logistic function $cx(1 - x)$ has an attracting or a repelling fixed point at 0.

7. Use graphical analysis to sketch the graphs of $Q^2(x)$ and $Q^3(x)$ for $-2 \leq x \leq 2$, where $Q(x) = x^2 - 2$.

8. Use graphical analysis to sketch the graphs of $S^2(x)$ and $S^3(x)$ for $0 \leq x \leq \pi$, where $S(x) = \pi \sin x$. What can you say about the number of fixed points that S^n has?

9. Consider the function

$$D(x) = \begin{cases} 2x & \text{if } 0 \leq x < \frac{1}{2} \\ 2 - 2x & \text{if } \frac{1}{2} \leq x \leq 1 \end{cases}$$

Sketch the graph of D, D^2, and D^3 for $0 \leq x \leq 1$. What can you say about the number of fixed points of D^n?

10. Consider the function

$$B(x) = \begin{cases} 2x & \text{if } 0 \leq x < \frac{1}{2} \\ 2x - 1 & \text{if } \frac{1}{2} \leq x \leq 1 \end{cases}$$

Sketch the graphs of B, B^2, and B^3. How many fixed points does B^n have? Can you find the periodic points of period 2 and 3 explicitly? How about the period n points? (This is challenging!)

Chapter 4

The Quadratic Family

In this section we will investigate the dynamics of the family of quadratic functions $Q_c(x) = x^2 + c$, where c is a parameter. We investigate this family for many different values of c; that's why we use the subscript. Like our friend the logistic function, this family represents one of the simplest nonlinear (i.e., not of the form $ax + b$) functions, yet we will see that the dynamics of this family are extremely complicated. We will return to this family again and again in this book, particularly when we study Julia sets and the Mandelbrot set.

4.1   Escaping Orbits of the Quadratic Function

Our first goal is to use a combination of graphical analysis and computer experimentation to understand how the dynamics of $Q_c(x) = x^2 + c$ change as we vary the parameter c. Toward that end, we first attempt to determine the c-values for which Q_c has interesting dynamics.

Experiment 4.1 Use ITERATE1 to determine the set of x_0-values whose orbits do not escape to infinity for each c.

Outcome. When $c > \frac{1}{4}$, it appears that all orbits of Q_c tend to infinity. When $-2 \le c \le \frac{1}{4}$, there appears to be an interval of x_0-values that do not escape. And when $c < -2$, it again appears that all orbits escape (we will see later that this last observation is far from the truth).

Let's use graphical analysis to understand the results of this experiment. The graph of Q_c is a parabola, which opens up as depicted in Figure 4.1. Note that this graph assumes three different positions relative to the diagonal, depending on whether $c > \frac{1}{4}$, $c = \frac{1}{4}$, or $c < \frac{1}{4}$. When $c > \frac{1}{4}$, the graph lies above the diagonal; when $c = \frac{1}{4}$, the graph just touches the diagonal; and when $c < \frac{1}{4}$, the graph meets the diagonal in two distinct points.

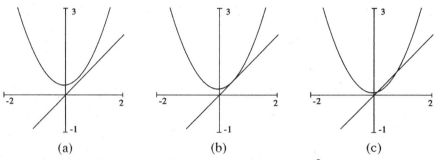

(a)   (b)   (c)

Figure 4.1 The graphs of $Q_c(x) = x^2 + c$
for (a) $c > \frac{1}{4}$, (b) $c = \frac{1}{4}$, and (c) $c < \frac{1}{4}$.

These facts are easy to verify analytically. The graph of Q_c meets $y = x$ whenever

$$x^2 + c = x$$

or

$$x^2 - x + c = 0$$

The roots of this quadratic equation are the x-values of points of intersection of the graph of Q_c and the diagonal. Using the quadratic formula, these points are easily determined to be

$$p = \frac{1 + \sqrt{1 - 4c}}{2}$$

$$q = \frac{1 - \sqrt{1 - 4c}}{2}$$

Note that p and q both depend on c, so we really should write $p(c)$ and $q(c)$ to indicate this dependence.

These numbers are real provided

$$1 - 4c \geq 0$$

Thus we see that there are points of intersection only when $c \leq \frac{1}{4}$. Note that $p = q$ when $c = \frac{1}{4}$ and that $q < p$ when $c < \frac{1}{4}$. These facts allow us to verify what we observed experimentally when $c > \frac{1}{4}$.

Exercise 4.2 Use graphical analysis to show that all orbits of Q_c tend to infinity when $c > \frac{1}{4}$.

Dynamically speaking, q and p are both fixed points for Q_c. Hence we see that a pair of fixed points are "born" as the parameter c decreases through $\frac{1}{4}$. This is an example of a *bifurcation*. Bifurcation means a change, a splitting apart or a division in two. In this example, we see that a fixed point for Q_c first appears when $c = \frac{1}{4}$ and then suddenly splits into two fixed points as c decreases. This is an example of what is known as a *saddle-node*, or *tangent*, *bifurcation*. We will see many other examples of this kind of bifurcation as we go along.

Where does the interval of x_0-values whose orbits do not escape when $-2 \leq c \leq \frac{1}{4}$ come from? Again graphical analysis yields the answer. Note first that $-p < q < p$ for all $c < \frac{1}{4}$. To see why this is true, we first note that

$$-\sqrt{1 - 4c} < \sqrt{1 - 4c}$$

Hence

$$1 - \sqrt{1 - 4c} < 1 + \sqrt{1 - 4c}$$

and also

$$-1 - \sqrt{1 - 4c} < 1 - \sqrt{1 - 4c}$$

Putting these two inequalities together yields

$$-1 - \sqrt{1 - 4c} < 1 - \sqrt{1 - 4c} < 1 + \sqrt{1 - 4c}$$

as long as $1 - 4c > 0$. If we divide each term in this inequality by 2, we get the desired result. Also note that $-p$ is an eventually fixed point, because $Q_c(-p) = p$, which is a fixed point.

Exercise 4.3 Use graphical analysis to show that if $x_0 > p$ or $x_0 < -p$, then the orbit of x_0 tends to infinity.

This exercise suggests that all the "interesting" dynamics of Q_c are confined to the interval $-p \leq x \leq p$ when $c \leq \frac{1}{4}$. We will denote this interval by I_c; note that the size of I_c depends on c since p does.

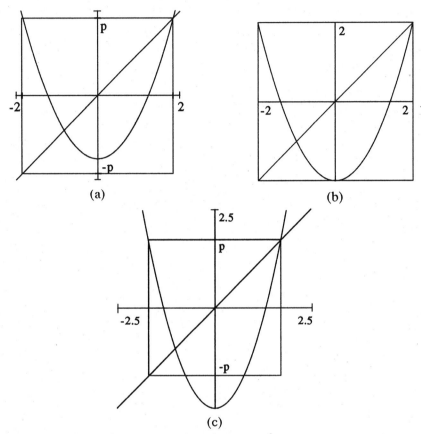

(a)

(b)

(c)

Figure 4.2 The graphs of Q_c for (a)$-2 < c < \frac{1}{4}$, (b)$c = -2$, and (c) $c < -2$.

Figure 4.2 depicts the graph of Q_c in three cases: $-2 < c < \frac{1}{4}$, $c = -2$, $c < -2$. In each case we have drawn a box centered at the origin with vertices at (p, p) and $(-p, -p)$. The portion of the graph of Q_c that is contained within this box is precisely the graph of Q_c on the interval I_c. Note that the lowest point on the graph of Q_c protrudes through the bottom of the box when $c < -2$. This is easily checked, since the lowest point on the graph is $(0, c)$ whereas the bottom of the box is given by

$$y = \frac{-1 - \sqrt{1 - 4c}}{2}$$

This y-value is just $-p$. Hence the lowest point on the graph protrudes through the bottom of the box provided

$$c < \frac{-1 - \sqrt{1 - 4c}}{2}$$

We may solve this inequality (subject to the requirement $c < \frac{1}{4}$) as follows. We have

$$2c + 1 < -\sqrt{1 - 4c}$$

The right hand side of this inequality is negative. Thus we must certainly have $2c + 1 < 0$, so that, in particular, $c < -\frac{1}{2}$. Taking absolute values of both sides yields

$$-2c - 1 = |2c + 1| > \sqrt{1 - 4c}$$

since $2c + 1$ is a negative number. Squaring both sides then gives

$$4c^2 + 4c + 1 > 1 - 4c$$

$$4c^2 + 8c > 0$$

$$4c(c + 2) > 0$$

This last product is positive if $c < -2$ or if $c > 0$. Thus, using the restriction that $c < -\frac{1}{2}$, we see that the lowest point on the graph of $Q_c(x) = x^2 + c$ does indeed protrude from the bottom of the box when $c < -2$.

Let's summarize what we have accomplished so far. We have found that all the interesting dynamics of Q_c occur when $c < \frac{1}{4}$. Moreover, for each such c, the interesting dynamics occur on the interval I_c given by $-p \leq x \leq p$ where

$$p = \frac{1 + \sqrt{1 - 4c}}{2}$$

4.2   The Interesting Orbits

Now we turn to a discussion of the orbits that do not tend to infinity, the most interesting orbits of the system. The importance of the box in Figure 4.2 is that it traps all the orbits of Q_c within I_c when $-2 \leq c \leq \frac{1}{4}$. All the vertical and horizontal lines associated with graphical analysis remain within this box for these parameter values. See Figure 4.3. This means that if our

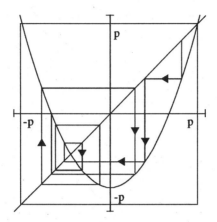

Figure 4.3 Orbits trapped in I_c for $-2 \leq c \leq \frac{1}{4}$.

initial seed x_0 is any point in I_c, then the entire orbit of x_0 is trapped forever in I_c. That is, all the interesting orbits of Q_c in this case lie in I_c.

When $c < -2$, there are points in I_c (such as 0) whose orbits escape from I_c and then tend to infinity. But there are many other points whose orbits remain in I_c. We return to this case later when we describe Cantor sets. For the remainder of this section, however, we deal with the dynamics of Q_c when $-2 \leq c \leq \frac{1}{4}$. To understand the dynamics of this function, let us begin with some numerical experiments.

Experiment 4.4 Modify ITERATE3 and/or ITERATE1 so that the new program does the following:
1. Accepts as input an initial seed x_0 and a parameter value c. You should choose x_0 between $-p$ and p; $x_0 = 0$ is always a good choice.
2. Displays the first 100 iterates of x_0 under Q_c in the interval $-p \leq x \leq p$.

Note that your program must compute the value of p that depends on c via the relation

$$p = \frac{1 + \sqrt{1 - 4c}}{2}$$

Use this modified program to observe the orbits of Q_c for a large number of c-values in the range $-2 \leq c \leq \frac{1}{4}$. Record your observations.

Outcome. There are a variety of different behaviors depending on c, much the same as we observed in our experiment in Chapter 2 with the logistic function $F(x) = cx(1-x)$. For $-\frac{3}{4} \leq c \leq \frac{1}{4}$, all orbits appear to be attracted to an attracting fixed point. For $-\frac{5}{4} \leq c < -\frac{3}{4}$, most orbits appear to be attracted to a period 2 cycle. As c decreases further, most orbits are attracted to a period 4 cycle, then a period 8 cycle, then a period 16 cycle, and so forth. There are many c-values for which no pattern may be discerned. Can you find c-values for which 0 is attracted to a cycle of period 3?

4.3   The Period-Doubling Bifurcation

Let's use graphical analysis to explain some of this behavior. We already know that all orbits tend to infinity when $c > \frac{1}{4}$. When $-\frac{3}{4} \leq c \leq \frac{1}{4}$, all orbits appear to tend toward an attracting fixed point. This is the point

$$q = \frac{1 - \sqrt{1 - 4c}}{2}$$

that we computed previously. This graphical analysis is depicted in Figure 4.4. You might also check this numerically using ITERATE1 or ITERATE3. In this case q is an attracting fixed point and the interval $-p < x < p$ is its basin of attraction.

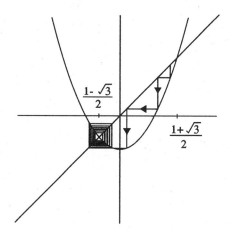

Figure 4.4 Graphical analysis of $Q_{-\frac{1}{2}}$.

When $c = -\frac{3}{4}$, another type of bifurcation occurs. This bifurcation is called a *period-doubling bifurcation*. According to the outcome of our preceding experiment, all orbits of points in I_c are attracted to a single fixed point when $-\frac{3}{4} < c < \frac{1}{4}$. (When c is close to $-\frac{3}{4}$ or to $\frac{1}{4}$, it takes a long time for orbits to "reach" the fixed point.) When c passes through $-\frac{3}{4}$, this suddenly changes and all orbits of points in I_c appear to be attracted to a cycle of period 2. When c is very close to $-\frac{3}{4}$, this cycle contains two points that are very nearly the same, but they spread apart as c decreases. You should experiment with a variety of c-values to observe this behavior.

Thus it appears that as c decreases through $-\frac{3}{4}$, the attracting fixed point q disappears and a new attracting cycle of period 2 is born. Actually, q has not disappeared, as we know from our previous computation of q; we have seen that q is a fixed point for all values of $c < \frac{1}{4}$. All that has happened is that q has changed from an attracting to a repelling fixed point. Meanwhile, at the same time, a period 2 attracting cycle is born.

All of this can be explained by graphical analysis. Figure 4.5 displays portions of the graphs of Q_c for c-values near $-\frac{3}{4}$, while Figure 4.7 displays the graphs of Q_c^2 for the same c-values. In Figure 4.5 we have also drawn a line that is perpendicular to the diagonal at (q, q). When $c = -\frac{3}{4}$, the graph of Q_c is tangent to this perpendicular line. As c passes through this value, the graph twists from one side of this perpendicular line to the other.

What does this mean dynamically? Figure 4.5 shows that when $c \geq -\frac{3}{4}$, q is attracting, but when $c < -\frac{3}{4}$, q becomes repelling. Note that points on nearby orbits oscillate from one side of q to the other. When $c < -\frac{3}{4}$, points near q move further away under Q_c^2. However, points that are far away from q tend to move closer under Q_c^2, as depicted in Figure 4.6. This means that somewhere in between there must be a point that moves neither farther away from nor closer to q, that is, a point fixed by Q_c^2. So somewhere in between there must be a cycle of period 2.

This is most easily seen by looking at the entire graph of Q_c^2, which is depicted in Figure 4.7. Note the birth of two new fixed points for Q_c^2 as c decreases through $-\frac{3}{4}$. Graphical analysis shows that these are attracting fixed points for Q_c^2. Since they are not fixed by Q_c, they must lie on an attracting cycle of period 2.

This is a typical period-doubling bifurcation. As a parameter varies, a given periodic orbit changes from attracting to repelling. Meanwhile, a new cycle of twice its period appears.

We can verify all of this analytically. The period 2 points are solutions to the equation

$$Q_c^2(x) = x$$

Working this out, we find

$$x^4 + 2cx^2 - x + c^2 + c = 0$$

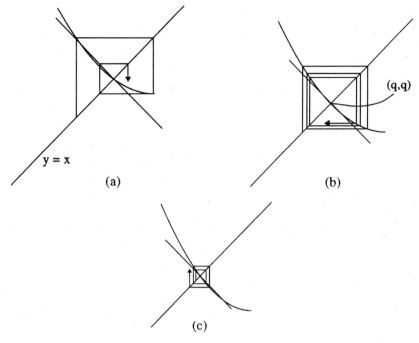

Figure 4.5 Portions of the graphs of Q_c
for (a) $c = -.65$, (b) $c = -.75$, and (c) $c = -.85$.

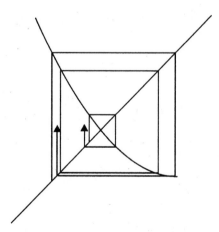

Figure 4.6 A cycle of period 2 for Q_c for $c = -.8$.

This is a fourth degree polynomial whose roots give the periodic points with period 2. Generally, such equations are difficult to solve, but here we already

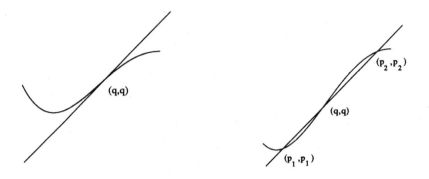

Figure 4.7 The graphs of Q_c^2 for c-values near $-\frac{3}{4}$.
The points p_1 and p_2 lie on a cycle of period 2.

know two roots. Indeed, both p and q are fixed points, so they certainly solve this equation. This means that $(x - p)$ and $(x - q)$ are factors of the left hand side of this equation. But remember, both p and q are roots of the quadratic equation

$$x^2 - x + c = 0$$

(see Section 4.1). So we may divide the fourth degree polynomial by this quadratic factor. This leaves us with a quadratic polynomial, whose roots are easily generated by the quadratic formula. We leave it to you to verify that this quadratic equation is given by

$$x^2 + x + c + 1 = 0$$

That is

$$\frac{x^4 + 2cx^2 - x + c^2 + c}{x^2 - x + c} = x^2 + x + c + 1$$

This quadratic equation has real roots when $c \leq -\frac{3}{4}$. The algebra here is complicated (remember, all the coefficients depend on c) so we leave the tedious details for you to work out. This is not exactly fun, but you can do it!

Experiment 4.5 Use ITERATE1 or ITERATE3 to investigate the period–doubling bifurcation that occurs in the logistic family $cx(1 - x)$ when $c = 3$. Can you find experimentally c-values for which similar bifurcations occur in the following functions?

  a. $F(x) = x^3 - cx$
  b. $E(x) = -c \exp x$
  c. $S(x) = d \sin x$

Outcome. You should use the graphs of these functions to help you find approximate c-values first, and then experiment.

Our earlier experiments showed that, as c decreases through $-\frac{5}{4}$, most orbits begin to be attracted to a period 4 cycle. How does this occur? Let's use graphical analysis again, this time applied to Q_c^2. Figure 4.8 shows the graph of Q_c^2 for various c-values. We have drawn a small box in each figure. Compare the portions of the graph of Q_c^2 with those of Q_c in Figure 4.2. Note the similarity. If we think exactly as we did before about what is happening dynamically inside this box, we expect Q_c^2 to undergo a period–doubling bifurcation, just as Q_c did before. That is, at $c = -\frac{5}{4}$, we expect the period 2 cycle to undergo a period–doubling bifurcation. At this c-value, a new cycle of period 4 is born, while the period 2 cycle becomes repelling.

This scenario continues as c decreases. The period 4 cycle eventually undergoes a period–doubling bifurcation, spawning a new attracting cycle of period 8, which then doubles and gives an orbit of period 16, and so forth. At each stage a cycle of period 2^n becomes repelling as an attracting cycle of period 2^{n+1} is born. Unfortunately, it is very hard to distinguish these orbits when n gets larger than 4.

This is called the *period–doubling route to chaos*, a phenomenon that has only recently been shown to occur in a great many dynamical systems.

Exercise 4.6 Use ITERATE1 or ITERATE3 to find experimentally the first few period–doubling bifurcations in the following families of functions.
  a. $S(x) = d \sin x$, $0 \le x \le \pi$
  b. $F(x) = cx(1 - x)$, $0 \le x \le 1$ when $c < 0$
  c. $F(x) = x^3 - cx$, $-2 \le x \le 2$

Exercise 4.7 Consider the family of cubic functions given by $T_c(x) = x^3 - cx$, where the parameter c satisfies $0 \le c \le 3$. What are the fixed points for T_c? Are they attracting or repelling? Use ITERATE1 or ITERATE3 to help you decide. Show that $\pm\sqrt{c-1}$ lies on a cycle of period 2 when $c > 1$. Hence a period–doubling bifurcation occurs when $c = 1$. Explain this using the graphs of T_c and T_c^2 for various c-values.

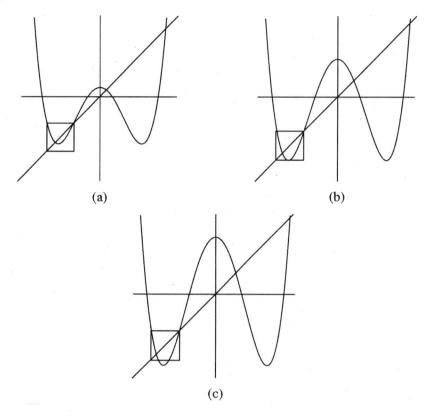

(a)   (b)

(c)

Figure 4.8 Compare the portions of the graphs of Q_c inside
the box to the graphs of Q_c in Figure 4.2.

4.4   The Chaotic Quadratic Function

Now let's turn our attention to another c-value with quite different dynamics, namely, $c = -2$. This is precisely the point at which our experiments
showed a dramatic change in the dynamics of Q_c. For $c = -2$, it appears
that a "typical" orbit fills up the interval $-2 \leq x \leq 2$ (recall that $p = 2$
when $c = -2$). We saw this in Problem 4 at the end of Chapter 2. This
will be one of our principal examples when we discuss the concept of chaos
later. Other periodic points seem hard to find (except for the repelling fixed
point $p = 2$). We ask if there are any other cycles for Q_{-2}. The surprising
answer is that there are infinitely many. The reason for this is exactly the
same as for $4x(1 - x)$ in the previous chapter. We may understand this by
using graphical analysis.

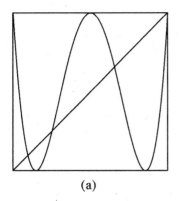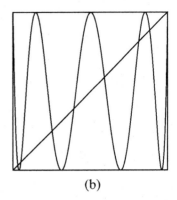

(a)   (b)

Figure 4.9 The graphs of Q_{-2}^2 and Q_{-2}^3.

Let's first draw the graphs of Q_{-2} and its iterates in the box centered at $(0,0)$ with vertices at $(-2,-2)$ and $(2,2)$. This is easily done using a graphing program or graphical analysis. Note that the graph of $Q_{-2} = x^2 - 2$ passes through the points $(-2,2)$ and $(2,2)$. The graph is a parabola whose vertex is located at $(0,-2)$. Also note that $Q_{-2}(\sqrt{2}) = 0$ and $Q_{-2}(-\sqrt{2}) = 0$. Therefore, $Q_{-2}^2(\pm\sqrt{2}) = -2$. Using graphical analysis, it follows that the graph of Q_{-2}^2 has 2 "valleys," as depicted in Figure 4.9. Arguing similarly, Q_{-2}^3 has four valleys, Q_{-2}^4 has eight valleys, and so on. Q_{-2}^n has 2^{n-1} valleys in this box. Consequently, the graph of Q_{-2}^n crosses the diagonal 2^n times in the interval $-2 \le x \le 2$. This means that Q_{-2}^n has at least 2^n fixed points in this interval. Not all these points are fixed by Q_{-2}; most have prime period n. In any event, we have shown that, by the time c has decreased from $\frac{1}{4}$ to -2, this quadratic family has developed infinitely many periodic points.

Notice the similarity between Q_{-2} and what we found in the last chapter for the function $4x(1 - x)$. The next section will show that there is a strong resemblance between all the members of the quadratic and the logistic families.

You should appreciate the power of this qualitative or geometric method. As we noted before, to find these points algebraically, we would need to solve the equation

$$Q_{-2}^n(x) - x = 0$$

This is a polynomial equation of degree 2^n whose roots we must find. If n is large, this is an impossible task. We have also seen that the computer does not help much either.

4.5 The Orbit Diagram

To summarize, we have seen that, as c decreases, the number of periodic points of Q_c increases and there are c-values with complicated orbit structure. Our next project allows us to put all this information together. We will construct the *orbit diagram* for Q_c. This is a diagram that plots the orbit of a particular point versus the parameter value c. We will choose 360 equally spaced values of c between $\frac{1}{4}$ and -2. We will then compute the orbit of 0 for each of these c-values and display each of these orbits on a different vertical line. The horizontal direction will represent the parameter c. So that we only see the ultimate behavior of the orbit of 0, we will not plot the first 50 iterates of the orbit. We will only plot the subsequent points on the orbit.

```
REM program ORBITDGM
FOR c = -2 TO 0.25 STEP 0.00625
   x = 0
   m = 160 * (c + 2)
   FOR i = 0 TO 200
      x = x * x + c
      IF i < 50 GOTO 11
      n = 75 * (2 - x)
      PSET (m,n)
   11 NEXT i
NEXT c
END
```

Figure 4.10 The program ORBITDGM.

Figure 4.10 gives a program called ORBITDGM, which produces the orbit diagram. The program steps through 360 different values of c between .25 and -2 via the statement

$$\text{FOR} \quad c = -2 \quad \text{TO} \quad .25 \quad \text{STEP} \quad .00625$$

Since $2.25/360 = .00625$, this FOR-NEXT statement selects 360 equally spaced c-values between .25 and -2, and then performs the necessary calculations for each c. The first 200 points on the orbit of 0 are calculated for

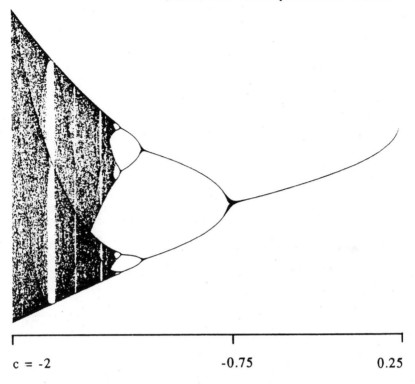

c = -2   -0.75   0.25

Figure 4.11 The output of ORBITDGM.

each c, but only the last 150 points are plotted. This allows us to see the eventual, or *asymptotic*, behavior of the orbit and is accomplished via the IF statement in the program.

Note that this program contains a nested loop. For each c-value, we first compute and then plot the orbit of 0 under Q_c, and then we increment c.

The output of this program is shown in Figure 4.11. Note that c increases from -2 to $.25$ as we move from left to right. This picture warrants some explanation. Remember that vertical lines contain the dynamics of a particular Q_c. Since only the last 150 points on the orbit of 0 are plotted, this means that when a small number of points appear on a given vertical line, we can assume that 0 has been attracted to an attracting cycle. With this in mind, we see clearly the period doublings from a fixed point to a cycle of period 2 to a cycle of period 4 and beyond. There is also a period 3 window, where 0 is attracted to an attracting cycle of period 3, and there are other c-values for which the orbit of 0 appears to be quite intricate. These c-values form the "chaotic regime," where orbits seem to move about randomly rather

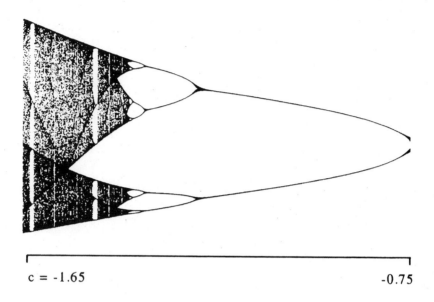

c = -1.65   -0.75

Figure 4.12 A magnification of the orbit diagram of Q_c.

than settling down to a fixed point or a periodic orbit. The fine structure in these areas is difficult to see at this resolution, so it is useful to magnify this picture as in the next projects.

Experiment 4.8 Modify ORBITDGM so that the program computes the orbit diagram for Q_c in the period-doubling regime, that is, for $-1.65 < c < -.75$. This orbit diagram is displayed in Figure 4.12. Here we see clearly the period-doubling bifurcations that give birth to the cycles of period 2, 4, and 8.

Experiment 4.9 Compute the orbit diagram for Q_c in the period 3 window $-1.8 < c < -1.75$ as well as in other areas in which the orbit of 0 appears to be attracted to an attracting cycle. Do you notice any similarities between these magnifications?

Outcome. In each case, it appears that the attracting orbit undergoes period-doubling as the parameter increases. The period 3 window is depicted

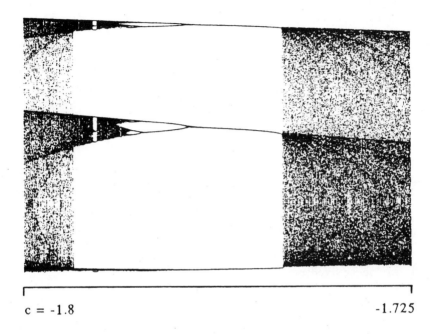

c = -1.8   -1.725

Figure 4.13 The period 3 window in the orbit diagram.
Note the period doublings here from 3 to 6 to 12 (barely visible).

in Figure 4.13.

Experiment 4.10 Compute the orbit diagrams corresponding to the initial
seed x_0 for the following families of functions. Do you notice any similarities
with the orbit diagram of Q_c?

   a. $F(x) = cx(1 - x)$, $0 \le x \le 1$, $1 \le c \le 4$; $x_0 = \frac{1}{2}$
   b. $S(x) = d \sin x$, $0 \le x \le \pi$, $0 \le d \le \pi$; $x_0 = \pi/2$

Outcome. Up to a change of scale, each of these orbit diagrams is identical!
It is one of the amazing facts of dynamics that simple functions like these
always tend to become chaotic in the same manner. Note the clearly visible period doubling sequence with which each family begins its evolution.
Note also the period 3 window as well as other regions in which there is an
attracting periodic orbit. Using the magnification discussed earlier, can you
determine the periods of the basic cycle in these windows?

Remark. You may wonder why we chose the value $x_0 = 0$ to plot the orbit diagram for $x^2 + c$ (or $x_0 = \frac{1}{2}$ for $cx(1 - x)$ or $x_0 = \pi/2$ for $d \sin x$). We will explain this choice much later when we discuss the Mandelbrot set. For now, we merely mention that you will get "essentially" the same orbit diagram no matter which initial seed x_0 you choose within the designated interval. Occasionally this will fail if, for example, you inadvertently choose an x_0 that itself lies on a repelling cycle. But this rarely occurs, as we have seen.

Further Exercises and Experiments

1. Compute the orbit diagram for $T_c(x) = x^3 - cx$ for $-2 \le x \le 2$ and $0 \le c \le 3$. Use the initial seed $x_0 = \sqrt{c/3}$.

2. Consider $T_3(x) = x^3 - 3x$.
   a. Show that the fixed points for T_3 are $0, 2, -2$.
   b. Are these fixed points repelling or attracting?
   c. Show that $\pm\sqrt{2}$ lies on a cycle of period 2 for T_3. Is this cycle attracting or repelling?
   d. Find all points x that satisfy $T_3(x) = 2$. Remember, $x = 2$ is one of them; there is only one other.
   e. Find all points x that satisfy $T_3(x) = -2$. There are only two such points.
   f. Use the information in a–e to sketch carefully the graph of T_3 for $-2 \le x \le 2$.
   g. Now sketch the graph of T_3^2 on this interval.
   h. Also sketch T_3^3.
   i. How many fixed points do you expect for $T_3, T_3^2, T_3^3, T_3^4, \ldots, T_3^n$?
   j. Can you find any of these using ITERATE1?

3. This experiment deals with the family of functions $C_c(x) = c \cos x$.
   a. Plot the orbit diagram for C_c where $-\pi \le x \le \pi$ and $0 \le c \le \pi$. Use $x_0 = 0$ as initial seed.
   b. Using magnifications of the orbit diagram, find (at least approximately) intervals of c-values for which there is a single attracting fixed point, an attracting period 2 cycle, an attracting period 4 cycle, and so forth. Is there a period 3 window?
   c. Something strange happens to the orbit diagram for $c > 2.97 \ldots$. Use graphical analysis to explain what you see.

Chapter 5

Iteration in the Complex Plane

We turn now to one of the most fascinating topics in dynamics, iteration of functions of a complex rather than a real variable. Since many physical processes depend on more than one variable, it is natural to consider dynamical systems in more than one dimension. Also, the dynamics of these functions yield computer graphics images of great beauty and considerable contemporary interest. Before turning to the computer experiments, we need to review some facts about complex numbers.

5.1   Complex Numbers

Recall that the imaginary number i satisfies $i^2 = -1$, that is, $i = \sqrt{-1}$. Clearly, i is not a real number; rather, it is an example of a complex number. A complex number is a number of the form $z = x + iy$, where x and y are real numbers. Sometimes we write $x + yi$ instead of $x + iy$. For example, $2 + 3i$, $7 = 7 + 0i$, and $3i = 0 + 3i$ are all complex numbers. The number x is called the real part of $x + iy$, and y is the imaginary part.

Complex numbers are clearly not used for counting. They arise primarily because of the need to solve certain equations that have no real numbers as solutions. For example, the equation

$$x^2 + 1 = 0$$

clearly has no real solutions, but it does have the complex roots $z = \pm i$. Similarly, the equation

$$x^2 - 4x + 5 = 0$$

has, using the quadratic formula, two complex roots, $2 \pm i$.

Another reason for the usefulness of complex numbers is geometric, since complex numbers may be plotted in the plane in the natural way, with $x + iy$ plotted at the point with coordinates (x, y). Just as we may name any point on the number line using a real number, we may name any point in the plane using a complex number. So i is plotted at $(0, 1)$, whereas real numbers like $7 = 7 + 0i$ are plotted along the x-axis. In particular, the origin corresponds to the complex number $0 = 0 + 0i$. The *modulus* $|x + iy|$ is defined to be the distance from $x + iy$ to the origin. That is,

$$|x + iy| = \sqrt{x^2 + y^2}$$

Consequently,

$$|2 + 3i| = \sqrt{4 + 9} = \sqrt{13} \quad \text{and} \quad |i| = 1$$

Note that the modulus of a real number is just its absolute value, which is the reason for using the same notation for these two concepts.

We may operate algebraically with complex numbers in the natural way. For example, we add two complex numbers by adding their respective real parts and imaginary parts. That is, the sum of $2 + 3i$ and $4 + 6i$ is $(2 + 4) + (3 + 6)i = 6 + 9i$. The product of the two complex numbers $x + iy$ and $u + iv$ is defined using the distributive property of multiplication.

$$(x + iy) \cdot (u + iv) = xu + i^2 yv + ixv + iyu$$
$$= (xu - yv) + i(xv + yu)$$

Thus

$$(2 + 3i) \cdot (4 + 6i) = 8 - 18 + i(12 + 12)$$
$$= -10 + 24i$$

and

$$(x + iy)(x + iy) = x^2 - y^2 + i(2xy)$$

Addition and multiplication of complex numbers obey all of the usual rules of algebra, including the commutative, associative, and distributive laws. Without digressing to prove this here, we leave it as an exercise to check these rules by computing the following sums and products in several different ways.

Exercise 5.1 Compute the following:

a. $(3 + 2i) + (2 - i)$
b. $(6i) + (3 + 2i) + 5$
c. $(6i) \cdot (2 - 3i)$
d. $(1 + 4i) \cdot (7 + 3i)$
e. $(1 + 2i) \cdot ((3 + 2i) + (1 + 7i))$
f. $(1 + 2i) \cdot (1 + 4i)$
g. $(1 + 4i) \cdot (1 + 2i)$
h. $4i + 4i + 4i$
i. $4i \cdot (4i + 4i)$
j. $(4i + 4i) \cdot 4i$

Geometrically, addition of two complex numbers may be interpreted as follows. To add the complex numbers z and w, we first draw an arrow from 0 to z and from 0 to w. To find the complex number $z + w$, we simply translate the arrow terminating at w so that it now begins at the tip of z. The endpoint of this new arrow is $z + w$. We can similarly obtain $z + w$ by translating the arrow to z so that it begins at w. See Figure 5.1.

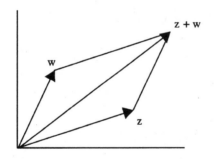

Figure 5.1 Addition of complex numbers.

Note that this geometric interpretation of addition immediately gives us the *triangle inequality*

$$|z + w| \leq |z| + |w|$$

This important inequality may be interpreted as saying the side of the triangle in Figure 5.1 stretching from 0 to $z + w$ has length that is no more than the sum of the sides from 0 to z and from z to $z + w$. Stated this way, the inequality is obvious.

There are other versions of this important inequality that we will use over and over. For example, if we replace w by $y - z$, we get

$$|y| = |z + y - z| \le |z| + |y - z|$$

using the usual triangle inequality applied to z and $y - z$. If we bring $|z|$ to the other side of this equation, we find

$$|y - z| \ge |y| - |z|$$

If we now replace z by $-z$, we find

$$|y + z| \ge |y| - |-z|$$

Therefore, since $|-z| = |z|$, we have

$$|y + z| \ge |y| - |z|$$

We will use these versions of the triangle inequality several times in the next few chapters.

5.2   The Program ITERATE4

Since multiplication of complex numbers makes sense, we can consider an old friend, the squaring function, as a complex function. Let $T(z) = z^2$, where $z = x + iy$ is a complex number. In terms of the real and imaginary parts of z, T is given by

$$T(x + iy) = x^2 - y^2 + i(2xy)$$

Thus the real part of $T(x + iy)$ is $x^2 - y^2$, and the imaginary part is $2xy$. This is a perfectly good function: When we apply T to a complex number, we get a new complex number, namely, z^2. We may ask what happens when we iterate T. The orbit of any complex number under this function is a collection of points in the complex plane rather than on the real line, as before. We may plot these points as we did before, remembering that there are now two coordinates to plot instead of just one.

Figure 5.2 gives a BASIC program called ITERATE4 to iterate the squaring function T. The program accepts as inputs the real and imaginary parts of a complex number $x_0 + iy_0$ and then plots in the complex plane the first

```
REM  program ITERATE4
INPUT  "x0"; x0
INPUT  "y0"; y0
FOR   j = 1  TO  100
    m = (x0 + 2) * 75
    n = (2 - y0) * 75
    PSET (m,n)
        FOR  i = 1  TO  1000
        NEXT  i
        PSET (m,n), 30
        FOR  i = 1  TO  100
        NEXT  i
    x1 = x0 * x0 - y0 * y0
    y1 = 2 * x0 * y0
    x0 = x1
    y0 = y1
NEXT  j
END
```

Figure 5.2 The program ITERATE4.

100 points on the orbit of $x_0 + iy_0$. We plot only points that lie within the square

$$|x| \leq 2$$
$$|y| \leq 2$$

If a point on the orbit lies outside this square, we do not plot it. In fact, when you compute certain orbits using this program, you will note that you very quickly get overflow messages from the computer. We will see why this happens shortly. As always, we must convert points in the complex plane to their screen coordinates. In this example, we have transformed the square $-2 \leq x_0 \leq 2$, $-2 \leq y_0 \leq 2$ in the complex plane into a 300×300 square on the screen. The screen coordinates (m, n) are given by the transformation

$$m = \frac{x_0 + 2}{4} * 300 = (x_0 + 2) * 75$$
$$n = \frac{2 - y_0}{4} * 300 = (2 - y_0) * 75$$

This transformation is depicted in Figure 5.3. Note that $x_0 + iy_0 = 0 + i0$ is transformed to $(150, 150)$, the center of the screen, whereas $-2 + i2$ is transformed to $(0, 0)$, the upper left–hand corner of the screen.

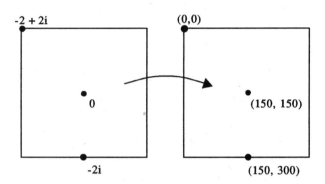

Figure 5.3 Transforming the complex plane to screen coordinates.

Transforming squares in the complex plane to squares on the screen is not too much different from the process we used to plot points along a line in Chapter 2. Often, because of the simplicity of the dynamical system we study, the squares of interest will be centered at the origin. But this need not be the case. Try your hand at constructing a transformation that takes the given square in the next exercise onto a 300×300 grid on the screen.

Exercise 5.2 Convert the following squares in the complex plane into screen coordinates by exhibiting the required transformation:

    a. $|x_0| \leq 3$, $|y_0| \leq 3$
    b. $-1 \leq x_0 \leq 3$, $1 \leq y_0 \leq 5$
    c. $-5 \leq x_0 \leq 0$, $|y_0| \leq 2.5$

We have included in ITERATE4 several empty loops of the form

        FOR $I = 1$ to 1000
        NEXT I

as pauses, which allow us to track points on the orbit as they are plotted; in this loop, the computer simply counts to 1000 before moving on to compute the next point on the orbit. As before, the statement

        PSET$(M, N), 30$

simply erases the point that had been plotted previously. This allows us to see the orbit clearly as it is being plotted.

Experiment 5.3 Use ITERATE4 to describe the fate of any orbit of T.

Outcome. It appears that any initial point $x_0 + iy_0$ with $|x_0 + iy_0| < 1$ has orbit that converges to 0, whereas if $|x_0 + iy_0| > 1$, the orbit tends to infinity. This is apparent because of the overflow messages you will receive on your screen. As we saw before, when an orbit converges to 0, it does so fairly quickly. Therefore, it appears that only a few points on the orbit are plotted before the orbit reaches 0.

Note that the outcome of this experiment agrees with what we found when we iterated T in the real case. The fact that orbits of points that satisfy $|x_0 + iy_0| < 1$ tend to 0 should come as no surprise. Indeed, we may compute

$$
\begin{aligned}
|T(x_0 + iy_0)| &= |(x_0^2 - y_0^2) + i(2x_0y_0)| \\
&= \sqrt{(x_0^2 - y_0^2)^2 + (2x_0y_0)^2} \\
&= \sqrt{x_0^4 + y_0^4 - 2x_0^2y_0^2 + 4x_0^2y_0^2} \\
&= \sqrt{x_0^4 + 2x_0^2y_0^2 + y_0^4} \\
&= \sqrt{(x_0^2 + y_0^2)^2} \\
&= x_0^2 + y_0^2 \\
&= |x_0 + iy_0|^2 \\
&< |x_0 + iy_0|
\end{aligned}
$$

This last inequality follows from the fact that $|x_0 + iy_0| < 1$, so the square of this number is smaller than itself.

The inequality $|T(x_0 + iy_0)| < |x_0 + iy_0|$ tells us that points that satisfy $|x_0 + iy_0| < 1$ move closer to 0 under one iteration of T. As we continue to iterate, this fact shows that the orbit continues to move closer to 0, which is a fixed point. The previous experiment shows that some orbits may circulate about 0 as they approach 0. Similar arguments show that points with $|x_0 + iy_0| > 1$ have orbits that tend to infinity under iteration of T.

So we have two predominant behaviors for orbits of points under T. Points with modulus less than 1 tend to 0, which is a fixed point, under iteration. Points with modulus larger than 1 escape to infinity. In between are points with modulus exactly 1. These points lie on the circle of radius 1 in the complex plane. This set is what is known as the *Julia set* of T. We do not describe the behavior of orbits of points in the Julia set now, but we return to this question later when we discuss chaos. We also see later that most Julia sets are much more spectacular, geometrically speaking, than the circle of radius 1.

The preceding argument can be easily adapted to show that if $|x_0 + iy_0| = 1$, then $|T(x_0 + iy_0)| = 1$ as well. That is, if a point starts on the Julia set, its orbit necessarily remains there forever. Because of this, we say that the Julia set is *invariant* under T. Even though we know that orbits are trapped on the Julia set, it is difficult to find an orbit that actually does stay there (except for ± 1 or $\pm i$). Try the following experiment to see this.

Exercise 5.4 Use ITERATE4 to compute the orbits of the following points under $T(z) = z^2$. Note that each of these points has modulus 1.

   a. $.6 + .8i$

   b. $(1 + i)/\sqrt{2}$

Outcome. Each of these points lies on the circle of radius 1 (compute the modulus to check this). However, when computed using ITERATE4, their orbits eventually fall off the circle and tend either to 0 or to infinity.

The culprit responsible for this is again round-off error. Small errors made in computing the orbit throws us off the Julia set. Once this occurs, we know what must happen — the orbit tends to 0 or to infinity. This foreshadows something we will see later — the Julia set is precisely the set of unstable orbits of a complex dynamical system.

Project 5.5 Modify ITERATE4 so that the new program circles each point as it is plotted, then a moment later erases the circle. This can be accomplished by replacing the five lines after the initial PSET command with

$$\text{CIRCLE}(m, n), 3$$
$$\text{FOR } i = 1 \text{ to } 1000$$
$$\text{NEXT } i$$
$$\text{CIRCLE}(m, n), 3, 30$$

To summarize, we have seen that, for the complex squaring function, a typical orbit falls into one of three categories.

   1. The orbit is attracted to 0.

   2. The orbit escapes to infinity.

   3. The orbit does neither of the above but rather remains forever on the Julia set.

In this sense, the Julia set is the boundary between the escaping orbits and those that never escape. We will see that the Julia set generally contains all the complicated orbits that we saw in the first four chapters. Moreover, for functions other than the squaring function, we will see that the Julia set is often quite complicated itself.

5.3 The Julia Set

The Julia set of a complex function is named for the French mathematician Gaston Julia, who discovered many of the basic properties of the set that bears his name in the early twentieth century. A precise definition of the Julia set of a polynomial is that it is the boundary of the set of points that escape to infinity. This means that a point in the Julia set has an orbit that does not escape to infinity, but arbitrarily nearby there are points whose orbits do escape.

Let's begin our study of Julia sets by modifying the program ITERATE4 so that the new program computes orbits for any quadratic polynomial of the form $z^2 + c$. Here both z and c will be complex numbers.

Project 5.6 Modify ITERATE4 to compute orbits of $Q_c(z) = z^2 + c$. Your new program should accept as input:
1. Any desired value of the constant $c = c_1 + ic_2$.
2. Any initial point $z_0 = x_0 + iy_0$.
3. Any desired number of iterates.
Call this new program ITERATE5. We use this program often in later sections.

Experiment 5.7 Use ITERATE5 to compute some of the orbits of the polynomial $Q_{-1}(z) = z^2 - 1$. Here $z = x + iy$ is a complex variable, so

$$Q_{-1}(x + iy) = x^2 - y^2 - 1 + i(2xy)$$

in the complex plane.

Outcome. As we saw for the real polynomial $x^2 - 1$, there are two predominant behaviors: Either orbits escape, or else they tend to an attracting cycle of period 2 that oscillates between 0 and -1. The points whose orbits neither escape nor tend to the cycle comprise the Julia set, though it is difficult to picture what this set looks like right now.

Experiment 5.8 Use ITERATE5 to compute orbits of the complex polynomial $Q_{-2}(z) = z^2 - 2$.

Outcome. It appears that all orbits except those between -2 and $+2$ on the x-axis tend to infinity.

Project 5.9 Modify ITERATE5 so that
1. Iteration stops if a point on the orbit has modulus larger than 10.
2. Instead of simply plotting successive points on the orbit, the computer draws straight–line segments connecting successive points on the orbit. This gives a different and more geometric method of viewing orbits in the complex plane.

Experiment 5.10 Use ITERATE5 to compute some of the orbits of $Q(z) = z^2 + .37 + .16i$.

Outcome. Some of the orbits that do not escape appear to "fill out" circles around a fixed point. What is this fixed point?

Further Exercises and Experiments

1. Why is it true that if $|x_0 + iy_0| = 1$, then $(x_0 + iy_0)^2$ also has modulus 1?

2. Write a formula for the real and imaginary parts of $(x+iy)^3$ and $(x+iy)^4$.

3. Use the previous exercise to modify ITERATE4 to compute the orbits of the following.
   a. $F(z) = z^3$
   b. $F(z) = z^4$

4. Each of the following complex functions is of the form $F(z) = cz$, where c is a complex constant. Use ITERATE4 to compute a variety of orbits for each function. Record what you see. Can you draw any conclusions from these observations? Use more c-values to help you decide which functions have all nonzero orbits tending to infinity and which have all orbits tending to 0.
   a. $c = \frac{1}{2}$
   b. $c = 2$
   c. $c = i$
   d. $c = 1 + i$
   e. $c = .9i$
   f. $c = 1.1i$
   g. $c = .6 + .8i$
   h. $c = \frac{5}{13} + \frac{12}{13}i$

Chapter 6

The Julia Set: Basin Boundaries

In this chapter we begin a detailed study of the geometry of the Julia set. For simplicity, and also to tie in with our previous work, we concentrate on quadratic functions of the form

$$Q_c(z) = z^2 + c$$

where c is a complex parameter. That is, $c = c_1 + ic_2$, where c_1 and c_2 are real numbers. In this section we present an algorithm to compute the Julia set of Q_c that works well when Q_c has an attracting periodic orbit. As we have seen before, it is not always true that Q_c has an attracting cycle. Thus the algorithm we present here will fail for many c-values. However, in the next chapter we present a different algorithm that works better when there is no attracting cycle present. Also, it is difficult to predict ahead of time which c-values will yield a dynamical system that has an attracting cycle. Later, when we describe the Mandelbrot set, we will find an efficient way to accomplish this.

In this chapter we consider only c values with $|c| \leq 2$, for reasons that will become apparent in Chapter 8. Since

$$|c| = \sqrt{c_1^2 + c_2^2}$$

we are therefore looking at those c-values that lie on or inside the circle of radius 2 in the c-plane. Recall from Chapter 4 that, when c is real, all the interesting dynamics occur on the interval $-2 \leq c \leq \frac{1}{4}$, so our experiments in this chapter include this case.

6.1   Escaping Orbits

Now let's return to the computer graphics. In case Q_c has an attracting periodic orbit, there is a collection of points that are attracted to the cycle. These are points that lie in what we have called the basin of attraction of the cycle. As we saw in the previous section, for $T(z) = z^2$, there are also points whose orbits tend to infinity. In between there are points that do neither; these points lie on the boundary between the basin of attraction and the escaping points. This is the set that is called the *Julia set*. So our first method for plotting the Julia set is to color points white if their orbits escape and to color them black if they do not. The boundary between these two regions is then the Julia set.

How do we decide if an orbit tends to infinity? For quadratic functions of the form $z^2 + c$ with $|c| \leq 2$, this is easy to determine. The rule of thumb is: If any point on the orbit of z_0 lies outside the circle of radius 2, then the entire orbit escapes to infinity. So, to see if the orbit of z_0 escapes, all we need do is check whether any point on the orbit ever has modulus greater than 2; if this is the case, then z_0 lies on an escaping orbit. To see why this is true, recall that we are considering only c-values with $|c| \leq 2$. If $|z| > 2$ and $|c| \leq 2$, then we have

$$|Q_c(z)| = |z^2 + c| \geq |z|^2 - |c|$$

by one of our versions of the triangle inequality discussed in Chapter 5. But

$$|z|^2 - |c| > |z|^2 - |z| = (|z| - 1)|z|$$

since $|z| > |c|$, and therefore $-|c| > -|z|$. Now

$$|z| - 1 > 1$$

since $|z| > 2$. Therefore, we may write

$$|z| - 1 = 1 + \ell$$

for some number $\ell > 0$. Thus we have

$$|Q_c(z)| > (1 + \ell)|z|$$

In particular,

$$|Q_c(z)| > |z|$$

This means, of course, that $Q_c(z)$ is farther from the origin than z is. Now we may also apply this line of reasoning to $Q_c^2(z) = Q_c(Q_c(z))$, since $|Q_c(z)| > |z| > 2$. We find

$$|Q_c^2(z)| = |Q_c(Q_c(z))| > (1 + \ell)|Q_c(z)|$$
$$> (1 + \ell)^2|z|$$

Thus $Q_c^2(z)$ is even further from the origin than $Q_c(z)$. Continuing in this fashion, we find

$$|Q_c^n(z)| > (1 + \ell)^n|z|$$

Now recall that $\ell > 0$, so that $1 + \ell > 1$. Hence the numbers $(1 + \ell)^n$ grow as n gets large. We have

$$(1 + \ell)^n \to \infty$$

as $n \to \infty$, so it follows that

$$|Q_c^n(z)| \to \infty$$

as $n \to \infty$. That is, the orbit of z escapes to infinity.

This then gives us the test to decide if an orbit tends to infinity: If any point on the orbit has modulus that exceeds 2, then we know that the orbit must ultimately tend to infinity.

6.2 The Program JULIA1

We capitalize on this last fact in the program JULIA1 in Figure 6.1. This program accepts as input a complex parameter $c = c_1 + ic_2$. It then displays in black the set of points within the square $|x|, |y| \leq 2$ whose orbit has not escaped beyond the circle of radius 2 centered at the origin before the twentieth iteration of Q_c. Remember, though, that this test works only for the function $Q_c(z) = z^2 + c$ when $|c| \leq 2$.

More precisely, the algorithm that we use to produce the Julia sets of Q_c is given by the following:

1. Input c_1 and c_2.
2. Select a 200×200 grid in the plane.

```
REM  program JULIA1
INPUT  "c1"; c1
INPUT  "c2"; c2
CLS
FOR  m = 0 TO  200
        x0 = -2 + m/50
        FOR  n = 0 TO  100
        y0 = 2 - n/50
        x = x0
        y = y0
        FOR  i = 1 TO  20
                x1 = x * x - y * y + c1
                y1 = 2 * x * y + c2
                x = x1
                y = y1
                z = x * x + y * y
                IF  z > 4 THEN GOTO  10
                NEXT  i
        PSET (m,n)
        PSET (200 - m, 200 - n)
10 NEXT  n
NEXT  m
END
```

Figure 6.1 The program JULIA1.

3. For each point z_0 in this grid, compute the first 20 points on the orbit of z_0. Check at each stage of the iteration whether the corresponding point lies outside the circle of radius 2.

4. If any point on the orbit lies outside of the circle of radius 2, then stop iterating and color the original point z_0 white.

5. If all 20 points on the orbit of z_0 lie inside the circle of radius 2, then color the original point z_0 black.

Several comments are in order. This program takes quite a long time to run. We are computing up to 20 iterations of Q_c on each point in a 200×200 grid. That means we must compute up to 800,000 iterations of Q_c in order to draw the set! Each iteration in turn involves a number of additions and multiplications, so this is a lot of computations for a personal computer. When running this program, we suggest that you find something else to do, such as homework, to while away the time! You should also remember that, at the outset, the computer may only be plotting white points, so nothing may appear on your screen for quite a while.

As usual, we must change coordinates from the xy-coordinates of the complex plane to screen coordinates. The transformation that links xy-coordinates to screen coordinates is given by the statements

$$x = -2 + \frac{m}{50}$$
$$y = 2 - \frac{n}{50}$$

Here, screen coordinates are given by (m, n). Note that $1 \leq m, n \leq 200$, whereas $-2 \leq x, y \leq 2$. For each point (m, n) on the screen, the preceding transformation singles out a complex number of the form $x + iy$ whose orbit we then test. The test is performed by the IF-THEN statement, where we check whether or not $|z|^2 > 4$. It is quicker to test $|z|^2 > 4$ than to invoke the square root function and then test whether $|z| > 2$.

Note that there is a triple nested loop in the program. We cycle through the screen coordinates by first fixing the m-coordinate and selecting each n-coordinate. Note that the n-coordinates run only from 0 to 100. This is because we exploit a symmetry in the dynamical system. After one iteration, the orbits of z and $-z$ are the same; indeed, $Q_c(z) = Q_c(-z)$. Thus if we know that the orbit of z escapes, then so does the orbit of $-z$. Hence if we run our test on the point (m, n), we know the result as well for $(200 - m, 200 - n)$, which corresponds to $-z$. You must be careful when using this symmetry, however. It only works if your "window" in the complex plane is centered at the origin. Otherwise, in JULIA1, n must cycle from 0 to 200 and the statement

$$PSET(200 - M, 200 - N)$$

must be eliminated.

Experiment 6.1 Check the accuracy of your JULIA1 program by running it with the value $c = 0$, that is, $c_1 = 0 = c_2$.

Outcome. The output of this experiment should be a disk centered at (100,100) in screen coordinates, with radius 50 pixels. (This is a very time-consuming way to draw a disk!)

Experiment 6.2 Use JULIA1 to generate the set of points whose orbit does not escape under iteration of Q_c for

    a. $c = -1$
    b. $c = .3 - .4i$
    c. $c = .360284 + .100376i$
    d. $c = -.1 + .8i$

Outcome. The results of these experiments are depicted in Figure 6.2.

Remark. It is quite remarkable that pictures such as those in Figure 6.2 were seen for the first time only in the late 1970s and early 1980s! It is true that mathematicians knew what certain Julia sets looked like (as, for example, the simple Julia set of the squaring function), but it is also true that nobody really understood how incredibly different and complicated these pictures could be as the parameter c changed, until very fast computers became readily available. We invite you use the computer and JULIA1 to experiment by choosing different c-values and then plotting the Julia set. You may be the first human ever to see the Julia set you compute!

As we noted earlier, it is the boundary, or edge, of the black region computed by JULIA1 that is called the Julia set. The entire black region is sometimes called the *filled-in Julia set*.

You may rightly object that our use of only 20 iterations in JULIA1 to decide if an orbit escapes is wrong. Indeed, to be perfectly correct, we would have to check *all* points on the orbit, clearly an impossible task. It is a fact that, for many Julia sets of Q_c, 20 iterations suffice to give a very accurate picture. You may check this by running the preceding experiment with 40 or 100 iterations and comparing the results. Note that it is easy to change the program to accomplish this, but the resulting computer run takes hours.

6.3   Magnifying the Julia Set

There are a million different modifications of JULIA1 that can be used to produce interesting images. One natural program to write would enable the user to "zoom in" on a previously computed portion of the filled-in Julia set.

Figure 6.2 Julia sets for Q_c (a) $c = -1$,
(b) $c = .3 - .4i$, (c) $c = .360284 + .100376i$, (d) $c = -.1 + .8i$.

Project 6.3 Modify JULIA1 so that, after running JULIA1, the new program prompts the user to select the lower left vertex and the side length of a smaller square in the plane. The program should then recompute the portion of the filled–in Julia set within this square using a 200×200 grid. This finer grid yields a higher resolution magnification of a portion of the

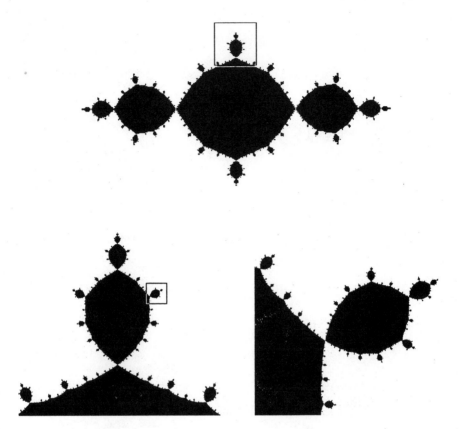

Figure 6.3 Magnifications of the Julia set of Q_{-1}.

filled-in Julia set.

Experiment 6.4 Use this modification of JULIA1 to compute successive enlargements of the filled-in Julia set of $Q_{-1}(z)$. You should always choose a portion of the previous picture that contains a portion of the Julia set (after all, all white or all black regions are not very interesting and they take quite a while to generate!) Remember that the symmetry must not be used when computing magnifications that are not centered at 0.

Outcome. See Figure 6.3. Note that successive enlargements of the Julia set of Q_{-1} reveal more and more "decorations" attached to the main black "bulbs" in the filled-in Julia set. You may need to increase the number of iterations to see this feature in satisfactory detail. The further we delve into the filled-in Julia set, the more decorations we see, but the pictures generated in smaller windows bear a strong resemblance to our original picture. This feature is called *self-similarity under magnification,* a basic property of *fractals,* a topic we take up later. See also Plates 1-3.

If your computer has a mouse, it would be useful to incorporate this feature into your program by allowing the user to select via the mouse the portion of a previous filled-in Julia set to be magnified.

The only difficulty in writing this program is in converting the new (x, y) coordinates to the 200×200 grid in screen coordinates. By now, you should have sufficient experience in transferring between these two coordinate systems, so we leave this computation as an exercise.

Project 6.5 Instead of using a 200×200 grid, better resolution can be obtained by using more points in the grid and more than 20 iterations to determine the filled in Julia set. You should be forewarned that an excessive number of iterations and a large grid size slow the computations down significantly, resulting in a program that may take hours or even days to run!

Project 6.6 Another interesting way to "capture" the filled-in Julia set is to use different colors to paint the set of escaping orbits, depending upon the time of escape. For example, if the orbit of a point z_0 escapes after n iterations (that is, the orbit has a point of modulus larger than 2 at iteration n), we might color the original point black if n is odd, or white if n is even. We can leave the original point white if it is in the filled-in Julia set. The result of this project is a succession of white and black "rings" that surround the filled-in Julia set.

Project 6.7 If your computer has color, striking images may be obtained by assigning different colors to different escape times, as determined under the previous project. For example, if your screen can display four other colors besides black, you might color points that escape before iteration 6 with one color, before iteration 11 with a second color, and so forth. This method allows us to get a feeling for the "dynamics" of Q_c near the Julia set. The resulting images give bands of points whose escape rates range from very quick to very slow. Of course, with more colors available, you can make the color bands finer, thus giving a more accurate picture of the dynamics as well as a more interesting image. See Plates 1-7.

Recall that in Chapter 4 we studied the saddle-node bifurcation. We

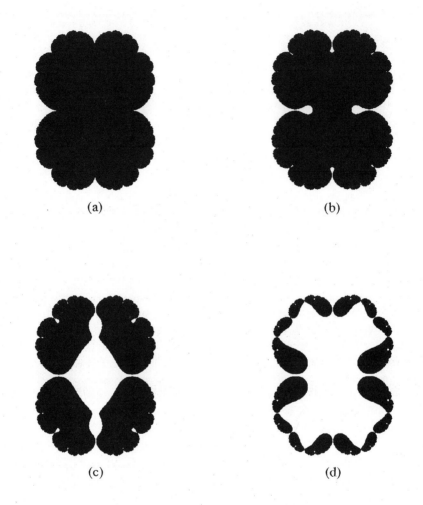

(a)   (b)

(c)   (d)

Figure 6.4 The saddle-node bifurcation for $z^2 + c$ where (a) $c = .25$, (b) $c = .2541$, (c) $c = .25448$, and (d) $c = .255$ using 50 iterations.

concentrated on the family of real functions

$$Q_c(x) = x^2 + c$$

with c near $\frac{1}{4}$. We saw that, for $c > \frac{1}{4}$, all points on the real line escaped to infinity, whereas if $c \leq \frac{1}{4}$, there was an interval of points that did not escape.

The next project allows you to see this same bifurcation in the complex plane.

Project 6.8 Compute the Julia set of $z^2 + c$ for a number of c-values in the range $.24 \leq c \leq .3$.

Outcome. You will see the filled-in Julia set "implode" as the parameter c increases through .25. See Figure 6.4. When $c > .25$, most points have orbits that escape under iteration. You may see this by fixing a c-value with $c > .25$ and then computing the filled-in Julia set using 20, 40, and then 60 iterations. Note how the black region grows smaller as the iteration count goes up. You should check that this does not happen if $0 \leq c \leq .25$.

A natural question is, What has happened to the filled-in Julia set as c increases through .25? We see later that this set has disintegrated into what is called *fractal dust* as we pass through the bifurcation point.

Project 6.9 If you have lots of computers at your disposal (as, for example, in a classroom or computer lab) or if you have lots of friends with computers, you can make a "movie" of the saddle-node bifurcation by computing many different filled-in Julia sets for c-values in the range $.24 \leq c \leq .3$.

Project 6.10 Investigate the period-doubling bifurcation at $c = -\frac{3}{4}$ in a similar fashion. What changes in the Julia set do you see as c decreases through $-\frac{3}{4}$?

Further Exercises and Experiments

1. Let $z = x + iy$ and $c = c_1 + ic_2$. Write out the formula for the complex logistic function $F(z) = cz(1 - z)$ in terms of x, y, c_1, and c_2.

2. The Julia set for the complex logistic function $F(z) = cz(1 - z)$, where c is now a complex parameter, may also be computed. However, we need to modify the algorithm somewhat. We need a test to check when an orbit escapes to infinity. Using the triangle inequality of Chapter 5, show that if

$$|z| > \frac{1}{|c|} + 1$$

then $|F(z)| > |z|$. Conclude that when z satisfies the preceding restriction, the orbit of z tends to infinity.

3. Use the results of exercises 1 and 2 to compute the filled-in Julia set of the complex logistic function for the following parameter values:

   a. $c = i$

   b. $c = 2$

   c. $c = 3.25$

   d. $c = 2 - i$

   e. $c = .6 + .8i$

   f. $c = \frac{5}{13} + \frac{12}{13}i$

4. Use the program JULIA1 and its modifications to investigate the Julia sets for the quadratic functions $Q_c(z) = z^2 + c$ where

   a. $c = -.48176 - .53165i$

   b. $c = -1.25$

   c. $c = -.39054 - .58679i$

   d. $c = -.11 + .67i$

5. Write out the real and imaginary parts of the expression $z^3 + c$, where both z and c are complex.

6. Let $T_c(z) = z^3 + c$. Use the triangle inequality to check that if $|z| > 2$ and $|c| \leq 2$, then $|T_c^n(z)| \to \infty$ as $n \to \infty$.

7. Use Exercises 5 and 6 to write a program analogous to JULIA1 that displays the Julia set of the cubic function $T_c(z) = z^3 + c$.

Chapter 7

The Julia Set: Other Algorithms

Our main goal in this chapter is to present alternative methods for displaying the Julia set of a polynomial function. The major method we discuss is the *backward iteration method.* This method has the advantage of working much more quickly than our previous method, but the pictures it displays are occasionally less precise. Other advantages include the fact that it produces an image of the Julia set itself rather than the filled-in Julia set and the fact that it works well when the previous method fails, notably when the Julia set is "fractal dust." Its disadvantage is the fact that we often find only the "skeleton" of the Julia set — many points that really lie in the Julia set are not found by this algorithm.

At the end of this chapter we outline a third method for computing Julia sets, the *boundary-scanning method.* This method combines many of the advantages of the previous two but takes even longer to run.

7.1   Polar Representation of a Complex Number

In the last two sections, we computed the orbit of a complex number z under iteration of the quadratic function $Q_c(z) = z^2 + c$. To do this, we needed to know how to square a complex number. To compute the backward orbit in this section, we will need to know how to "undo" this operation, that is, to compute the square root of a complex number.

To explain the operation of taking square roots, we first need to discuss the *polar representation* of a complex number. Suppose $z = x + iy$ is a complex number. In the Cartesian plane, z is located at the point with coordinates (x, y). We may equally well describe this point by giving its polar coordinates, namely, the modulus of z and the *polar angle*. The polar angle is the angle that the straight line connecting z to the origin makes with the positive x-axis. We measure this angle in radians and in the counterclockwise direction from the x-axis.

Recall that radian measure of angles is quite different from degrees: When we use radians, we are measuring the length of the arc of the circle of radius 1 that is subtended by that angle. Therefore, an angle of 360° is the same as an angle of 2π radians, since the circumference of a circle of radius 1 is exactly 2π. Similarly, angles of 180° or π radians are the same, and a right angle is an angle of $\pi/2$ radians.

Therefore, any point on the positive imaginary axis has polar angle $\pi/2$ and any point on the negative real axis has polar angle π. We agree to use negative angles if we measure polar angles in the clockwise direction, so points on the negative real axis also have polar angle $-\pi$, whereas points on the negative imaginary axis have polar angle $-\pi/2$ as well as $3\pi/2$.

We use the notation r for the modulus and θ for the polar angle. Thus, if $z = 1 + i$, then $r = \sqrt{2}$ and $\theta = \pi/4$. Similarly, if $z = 3i$, then $r = 3$ and $\theta = \pi/2$. Note that a given point may have many polar representations, for there is always an ambiguity in measuring the polar angle. For example, the point i has polar coordinates $r = 1$ and $\theta = \pi/2$. But an angle of $2\pi + \pi/2$ also determines the positive imaginary axis. So i also has the polar representation $r = 1$ and $\theta = 2\pi + \pi/2$. Clearly, any multiple of 2π can be added to the polar angle without changing the original point.

Exercise 7.1 Determine the modulus and the polar angle of each of the following complex numbers.

    a. $z = 2 + 2i$

    b. $z = -7$

    c. $z = -1 + i$

    d. $z = -4i$

    e. $z = 1 - \sqrt{3}i$

Given the polar representation of a complex number, we can determine its Cartesian coordinates using a little trigonometry. If z has polar coordinates r and θ, then we may draw the right triangle depicted in Figure 7.1. The hypotenuse of this triangle has length r, so the side opposite the angle θ has length $r \sin \theta$, and the adjacent side has length $r \cos \theta$. So $z = r \cos \theta + ir \sin \theta$.

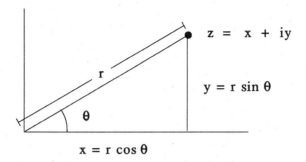

Figure 7.1 The polar representation of z.

Conversely, if we know the Cartesian coordinates of a complex number, then we may use the same triangle to determine the polar coordinates. If $z = x + iy$, then the triangle in Figure 7.1 has opposite side with length y and adjacent side of length x. Consequently,

$$\tan \theta = y/x$$
$$r = \sqrt{x^2 + y^2}$$

Therefore, the polar angle is given by

$$\theta = \arctan(y/x)$$

as long as $x > 0$, since the arctangent function takes only values between $-\pi/2$ and $\pi/2$. If $x < 0$,

$$\theta = \arctan(y/x) + \pi$$

For example, the point with modulus 2 and angle $\pi/6$ is given by $x + iy$, where

$$x = 2\cos(\pi/6) = \sqrt{3}$$
$$y = 2\sin(\pi/6) = 1$$

Using the polar representation of z, we can now compute the square roots of z. Just as in the real case, each complex number has two square roots. Suppose we are given the complex number $z = r\cos\theta + ir\sin\theta$. Then one of the square roots of z has modulus $\sqrt{r}$ and polar angle $\theta/2$, so that

$$\sqrt{z} = \sqrt{r}\,\cos(\theta/2) + i\sqrt{r}\sin(\theta/2)$$

As in the real case, the other square root of z is simply the negative of this complex number.

To check that all of this is true, we simply square the number $\sqrt{z}$ and use some trigonometry:

$$\sqrt{z} \cdot \sqrt{z} = \sqrt{r} \cdot \sqrt{r}(\cos^2(\theta/2) - \sin^2(\theta/2))$$
$$+ i\sqrt{r}\sqrt{r}(2\cos(\theta/2)\sin(\theta/2))$$

Now recall the two addition formulas from trigonometry:

$$\cos(A + B) = \cos A \cos B - \sin A \sin B$$

$$\sin(A + B) = \sin A \cos B + \cos A \sin B$$

If we apply these two formulas in the case where $A = B = \frac{\theta}{2}$, we find that

$$\sqrt{z} \cdot \sqrt{z} = r\cos(\theta/2 + \theta/2) + ir\sin(\theta/2 + \theta/2)$$
$$= r\cos\theta + ir\sin\theta$$

Thus, given a complex number z, we may find its two square roots by first computing its polar representation, obtaining the modulus r and polar angle θ of z. Then the two square roots of z are given by

$$w_1 = \sqrt{r}\cos(\theta/2) + i\sqrt{r}\sin(\theta/2)$$
$$w_2 = -\sqrt{r}\cos(\theta/2) - i\sqrt{r}\sin(\theta/2)$$

For example, to compute $\sqrt{i}$, we first find the polar representation of i. Clearly, this is $r = 1$ and $\theta = \pi/2$. Hence, one of the square roots of i has modulus $1 = \sqrt{1}$ and polar angle $\pi/4$. That is, one square root is

$$\sqrt{i} = \cos(\pi/4) + i\sin(\pi/4) = \frac{1}{\sqrt{2}} + \frac{i}{\sqrt{2}}$$

and the other is

$$\sqrt{i} = \cos(5\pi/4) + i\sin(5\pi/4) = -\frac{1}{\sqrt{2}} - \frac{i}{\sqrt{2}}$$

Exercise 7.2 Compute the square roots of each of the following complex numbers.

   a. $z = 4i$

b. $z = 1 + i$

c. $z = -1 + i$

d. $z = -7$

e. $z = -0.7$

The operation of taking the complex square root has an easy geometric interpretation: We simply take the square root of the modulus and halve the polar angle. See Figure 7.2.

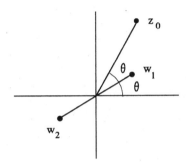

Figure 7.2 w_1 and w_2 are the two square roots of z_0.

7.2   The Squaring Function Again

Now let us see how the operation of taking square roots enables us to compute Julia sets. Recall first our discussion of the dynamics of $T(z) = z^2$ in Chapter 5. We saw then that all orbits of points that satisfy $|z| \neq 1$ tend to one of two places: If $|z| < 1$, the orbit tends to 0, but if $|z| > 1$, the orbit tends to infinity. Thus the circle of radius 1 is the boundary between points that are attracted to 0 and points whose orbits escape. This is the Julia set of T.

This suggests that we can find the Julia set of T by computing orbits backward, using the square root. If z_0 satisfies $|z_0| = r > 1$, then the two square roots of z_0 have modulus $\sqrt{r}$, and we have $1 < \sqrt{r} < r$. On the other hand, if $0 < |z_0| = r < 1$, then $r < \sqrt{r} < 1$. Notice that, in either case, $\sqrt{r}$ is closer to 1 than r is. This means no matter whether $r > 1$ or $r < 1$, the two square roots of z_0 lie nearer the circle of radius 1 than does z_0.

So, let's choose one of the two square roots of z_0; call this new point that we have chosen z_{-1}. Since $T(z_{-1}) = z_0$, we think of z_{-1} as being the first

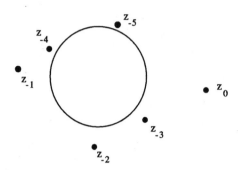

Figure 7.3 Backward orbits of $T(z) = z^2$.

point on one of the "backward" orbits of z_0. We may now repeat this process by choosing one of the two possible square roots of z_{-1}. Call this square root z_{-2}. We then continue in this fashion — selecting a particular square root and then taking one of its square roots. By our earlier reasoning, each time we select one of the two possible square roots, the point we have selected is closer to the Julia set. We thus see that the resulting backward orbit of the squaring function tends to the circle of radius 1, the Julia set of T.

In this process, at each stage one of two square roots must be chosen. We see shortly that it is best not to choose the "positive" or the "negative" square root each time. Rather, it is best to choose one of these two possibilities randomly. Figure 7.3 illustrates this procedure. Note that we have not always chosen the same square root in this figure and that the successive square roots $z_{-1}, z_{-2}, z_{-3}, \ldots$ of z_0 tend to the circle of radius 1.

The only point whose backward orbit fails to tend to the Julia set is 0: All other points have backward orbits that approach the Julia set.

7.3   The Program JULIA2

The preceding procedure works in general to produce a rough picture of the Julia set of $Q_c(z) = z^2 + c$. To compute a backward orbit of Q_c, we note that if

$$z^2 + c = w$$

then

$$z = \pm\sqrt{w - c}$$

The complex number $w - c$ has two square roots; these are the two preimages of w under Q_c.

Our earlier remarks suggest the following algorithm for computing the Julia set of Q_c.

1. Select any point w_0 in the complex plane.
2. Compute one of the two square roots of $w_0 - c$, choosing the positive or negative square root randomly. Let z_0 denote the value of this square root.
3. Replace w_0 by z_0.
4. Perform Steps 2 and 3 a total of 15,000 times, plotting at each stage the point z_0 in the plane. However, do not plot the first 50 points.

```
REM  program JULIA2
INPUT  "c1"; c1
INPUT  "c2"; c2
INPUT  "x0"; x0
INPUT  "y0"; y0
pi = 3.14159
CLS
      FOR  i = 1  TO  15000
      w0 = x0 - c1
      w1 = y0 - c2
      IF  w0 = 0  THEN   theta = pi/2
      IF  w0> 0 THEN   theta = ATN (w1 / w0)
      IF  w0 < 0 THEN   theta = pi + ATN (w1 / w0)
      r = SQR (w0 * w0 + w1 * w1)
            theta = theta/2 + INT (2 * RND) * pi
            r = SQR (r)
            x0 = r * COS (theta)
            y0 = r * SIN (theta)
            IF  i < 50  THEN GOTO  77
            m = (x0 + 2) * 300 / 4
            n =  ( 2 - y0) * 300 / 4
            PSET (m, n)
      77 NEXT  i
END
```

Figure 7.4 The program JULIA2.

In Figure 7.4 we have displayed the program JULIA2 that carries out this algorithm. This program accepts as input the parameter c and an initial seed z_0 and then computes 15,000 preimages of z_0. At each stage, the program randomly selects one of the two preimages. This is accomplished by using the RND statement from BASIC. Each time the program encounters the statement RND, it returns a random number between 0 and 1. Hence

$$2 * \text{RND}$$

returns a random number between 0 and 2. Therefore, the statement

$$\text{INT}\,(2 * \text{RND})$$

returns either the value 0 or 1 randomly, since the INT function rounds down to the nearest integer. Hence the statement

$$\text{THETA} = \text{THETA}/2 + \text{INT}(2 * \text{RND}) * \text{PI}$$

returns one of the two possible polar angles associated with the square root function; the new polar angle is randomly selected to be either $\theta/2$ or $\theta/2 + \pi$, depending upon whether

$$\text{INT}\,(2 * \text{RND}) = 0$$

or

$$\text{INT}\,(2 * \text{RND}) = 1$$

Note that in JULIA2 we do not plot the first 50 preimages, so that we only plot points on the backward orbit which are very close to the Julia set.

Experiment 7.3 Use JULIA2 to plot the Julia sets of Q_c for the following c-values.

    a. $c = -1$
    b. $c = -.4 - .6i$
    c. $c = -1.5$
    d. $c = -i$
    e. $c = -.8 + .4i$
    f. $c = .5$
    g. $c = 3$
    h. $c = 1 + i$
    i. $c = 2$

Outcome. The Julia sets for these c-values assume a variety of different shapes. See Figure 7.5. The Julia sets in parts (a)–(d) appear to consist of just one piece. For $c = -1$, the Julia set resembles the collection of decorated circles we saw before. In parts (c) and (d), the Julia set is a "dendrite," which seemingly bounds no region, as do the Julia sets in parts (a) and (b). In parts (e) and (f), the Julia sets appear to consist of many disjoint "islands" of points.

Experiment 7.3 shows that the backward iteration method yields a much quicker method of producing the Julia set. We need compute only 15,000 points on a single orbit rather than the 800,000 iterations that we needed for JULIA1. It has some disadvantages, however. The Julia set for Q_{-1} produced by this method does not seem to fill out the entire boundary displayed in Figure 6.2. Even if we increase the number of backward iterates in JULIA2, we seem to miss some of the points in the Julia set. Nevertheless, this method does yield a good approximation of the Julia set. The boundary scanning method discussed later remedies this defect.

JULIA2 produces pictures of certain Julia sets that are impossible to obtain by JULIA1, namely, those Julia sets that are not basin boundaries. This is the case in parts (e)–(h) of Experiment 7.3.

Experiment 7.4 For a given c-value, compute the Julia set of Q_c using a number of different initial seeds. Compare the pictures you obtain.

Outcome. The Julia set of Q_c generated by JULIA2 using a fixed c-value but different initial seeds are virtually identical. The only exception to this for the quadratic functions Q_c occurs for the initial seed 0 when $c = 0$. For $Q_0(z) = z^2$, the only preimages of 0 are 0 itself, so the backward orbit of 0 does not tend to the Julia set. Remarkably, this is the only exceptional case among all of the quadratic functions.

Experiment 7.5 Use JULIA2 to compute the Julia sets of Q_c for a variety of different c-values. How many different shapes can you find? Keep a "scrapbook" of as many different kinds of Julia sets and their corresponding c-values as you find. We see in the next chapter how the Mandelbrot set is a "table of contents" for your scrapbook.

7.4   Fractal Dust

There is one obvious qualitative difference among the Julia sets in Figure 7.5. In some cases, the Julia sets appear to form a single, connected piece,

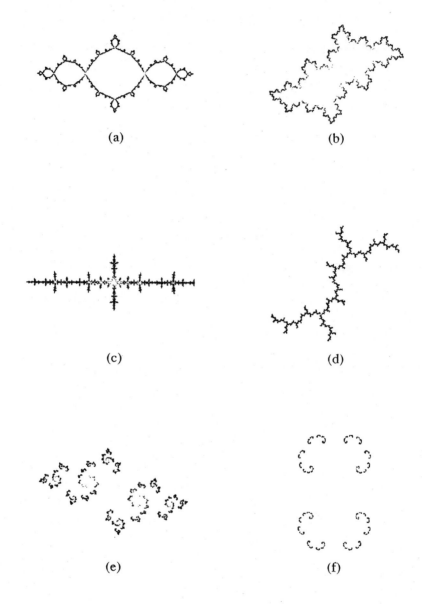

Figure 7.5 Julia sets of Q_c for (a) $c = -1$, (b) $c = -.4 - .6i$, (c) $c = -1.5$, (d) $c = -i$, (e) $c = -.8 + .4i$, and (f) $c = .5$.

whereas in other cases, the Julia set appears to consist of many pieces. From the pictures we have generated so far, it would appear that these latter Julia sets consist of finitely many isolated pieces. Actually, nothing could be further from the truth. The next project allows us to delve more deeply into the fractal nature of these Julia sets. We give a precise definition of the word *fractal* in Chapter 9.

Project 7.6 Modify JULIA2 so that the new program allows us to magnify a portion of the Julia set of Q_c. Your program should accept as input a small square within the original region and then recompute and display the backward orbit within this region. You may need to use more iterations to compute the magnified picture.

Experiment 7.7 Use this modified program to magnify various regions of the Julia set of Q_c when

   a. $c = -2.2$
   b. $c = -.8 - .8i$
   c. $c = -1 + i$
   d. $c = .5$

Outcome. In each case, if we magnify what appears to be an isolated piece of the Julia set, we see that this piece is itself made up of separate isolated chunks. Moreover, these smaller pieces bear a remarkable resemblance to the original Julia set! See Figure 7.6.

Again we see that the Julia sets of Q_c possess the property of self-similarity under magnification. This is one of the basic properties of a fractal. In Experiment 7.7, each successive magnification reveals that there are no isolated chunks of the Julia set. If we carry out this magnification over and over, we find that the Julia set consists of a "cloud" of points, each of which lies in a separate piece of the Julia set. These types of Julia sets therefore resemble *fractal dust*, although the technical term for this kind of structure is a *Cantor set*. We discuss Cantor sets in more detail in Chapter 9. For now we simply note that a Cantor set is *totally disconnected*. This means that if we take any two points in the Cantor set — for instance, z_0 and w_0 — there is always a closed curve that never meets the Cantor set and that surrounds z_0 but not w_0. See Figure 7.7. Thus we can always isolate two points in a totally disconnected set by means of a closed curve that does not intersect the set.

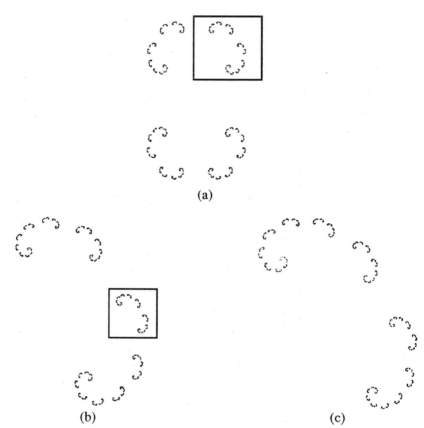

Figure 7.6 The Julia set of $Q_c(z) = z^2 + 0.5$ and several successive magnifications. Part (b) is a magnification of the box in Part (a). Note the similarity between (b) and (c).

At the opposite end of the spectrum are *connected* sets. A subset of the plane is connected if it is impossible to find a closed curve that is disjoint from the set and that separates the set into two disjoint pieces. For example, an interval is connected, but a pair of disjoint intervals is not a connected set.

Now let's turn to the question of what makes certain Julia sets totally disconnected and others apparently connected. The answer for this is simple and is illustrated vividly by the following experiment.

Experiment 7.8 Use the programs ITERATE5 and JULIA2 to compute the forward orbit of 0 for Q_c and the Julia set corresponding to the same

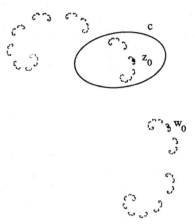

Figure 7.7 The curve c separates z_0 from w_0.

c-value. Use a variety of c-values, including

    a. $c = -1$
    b. $c = 0$
    c. $c = -2$
    d. $c = .2$
    e. $c = i$
    f. $c = .3$
    g. $c = .5$
    h. $c = 1$
    i. $c = 2 + i$
    j. $c = -3$

Outcome. When the orbit of 0 escapes to infinity, as in cases (f)–(j), the Julia set resembles fractal dust. When the orbit of 0 does not escape as in cases (a)–(e), the Julia set appears to be connected. You can use graphical analysis to check that the orbit does not escape in cases (a)–(d). Case (e) requires that you compute the orbit of 0 explicitly; do this and see that 0 is eventually periodic (and so never escapes).

The fact that the escape of the orbit of 0 governs whether or not the Julia set of Q_c is fractal dust is a true statement, which can be proved rigorously. This is the first instance in which we see that the orbit of 0 plays a special role in determining the dynamics of Q_c. We will see other instances of this as we go along, particularly when we discuss the Mandelbrot set in the next chapter.

Whenever the orbit of 0 escapes to infinity, the Julia set of Q_c is fractal dust, that is, is totally disconnected. On the other hand, if the orbit of 0 does not escape, the Julia set and the filled in Julia set are connected sets. The amazing part of this statement is the fact that there is no in between: Either the Julia set of Q_c consists of one piece or else it consists of infinitely many. For this reason, the point 0 is called a *critical point*. The orbit of 0 is called the *critical orbit*. The critical orbit thus plays a dominant role in determining what the Julia set of Q_c looks like. We make much use of this fact in the next chapter when we describe the Mandelbrot set.

This is a good time to return again to the saddle–node bifurcation, which we saw in the last chapter. Let's use JULIA2 to explore the Julia sets of Q_c for c-values near $\frac{1}{4}$.

Experiment 7.9 Use JULIA2 to compute the Julia sets of $Q_c(z) = z^2 + c$ for various real c-values near $c = \frac{1}{4}$.

Outcome. For $c \leq \frac{1}{4}$, the Julia set of Q_c appears to be a connected set — a curve. When $c > \frac{1}{4}$, this curve breaks up into infinitely many pieces. To be honest, it is not so clear that this change occurs precisely at .25, but there is clearly a change as c increases from .2 to .3. Also try $c = .4$ and $c = .5$.

The results of this experiment bear out what we described before.

Exercise 7.10 Use graphical analysis to check that the orbit of 0 under $Q_c(x) = x^2 + c$ escapes to infinity if $c > \frac{1}{4}$ but does not escape if $0 \leq c \leq \frac{1}{4}$.

We may also explain at this point another phenomenon associated to the saddle-node bifurcation. From the point of view of real dynamics, it appears that Q_c has two fixed points when $c < \frac{1}{4}$ and none when $c > \frac{1}{4}$. This was shown in Figure 4.1.

From the point of view of the complex plane, however, there is no sudden appearance of new fixed points at $c = \frac{1}{4}$. Indeed, if we solve the equation

$$x^2 + c = x$$

using the quadratic formula, we see that this equation has two complex solutions when $c > \frac{1}{4}$. This means that Q_c has two fixed points in the complex plane when $c > \frac{1}{4}$. These two fixed points simply come together and meet when $c = \frac{1}{4}$ and thereafter stay on the real line.

So where is the bifurcation in the complex plane? Our discussion of Julia sets provides the answer. We have seen that the Julia set of Q_c undergoes a major change when c decreases through $\frac{1}{4}$. For $c > \frac{1}{4}$, the Julia set of Q_c is totally disconnected, but when $-2 \le c \le \frac{1}{4}$, the Julia set is connected. The simple bifurcation at $c = \frac{1}{4}$ has the global effect of "gluing" together all the disjoint pieces of the Julia set at the same instant!

Experiment 7.11 Use JULIA2 to describe the global changes that occur in the Julia set at the period-doubling bifurcation at $c = -\frac{3}{4}$.

Outcome. The Julia set is a simple closed curve with no pinch points when $c > -\frac{3}{4}$. All of a sudden, this curve becomes pinched together at infinitely many points as c passes through $-\frac{3}{4}$.

7.5   The Boundary-Scanning Method

In this section we outline a third method for computing the Julia set of Q_c, which works very well when Q_c has an attracting cycle. The boundary-scanning method produces a very sharp image of the Julia set, not the filled-in Julia set, but it takes a long time to run on the computer. The idea behind this method comes from the very definition of the Julia set as the boundary between the set of points whose orbits escape and those whose orbits are attracted to a cycle. What we will do is color a pixel if the orbit corresponding to this point does not escape during the first 20 iterations but some adjacent pixel's orbit does escape.

The algorithm for the boundary-scanning method is as follows:

1. Select a 200×200 grid in the plane.
2. For each pixel (m, n) in this grid: Compute the first 20 points on the orbit corresponding to (m, n). If any of the points on this orbit have modulus greater than 2, color (m, n) white.
3. If the orbit corresponding to the pixel (m, n) does not escape, then compute the first 20 points on the orbits corresponding to the four pixels $(m + 1, n)$, $(m - 1, n)$, $(m, n + 1)$, and $(m, n - 1)$.
   a. If at least one of these additional orbits escapes, then color (m, n) black.
   b. If all four of these additional orbits do not escape, then color (m, n) white.

Project 7.12 Use the preceding algorithm to write a program called JULIA3 to compute the Julia sets of Q_c via the boundary scanning method. You might consider writing a subroutine to compute the various orbits required by this program using the GOSUB and RETURN statements. Remember to make use of symmetry to speed up the program when applicable.

Experiment 7.13 Use JULIA3 to recompute the Julia sets of Q_c when

    a. $c = -1$

    b. $c = -.1 + .8i$

    c. $c = .3 - .4i$

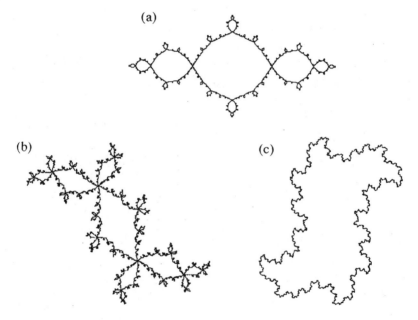

Figure 7.8 Julia sets for Q_c when (a) $c = -1$, (b) $c = -.1 + .8i$, and (c) $c = .3 - .4i$ computed using JULIA3.

Outcome. Some of the images generated by the boundary-scanning method are shown in Figure 7.8. Note that the points that are "missing" from the Julia set when $c = -1$ in Figure 7.5 are now included.

Chapter 8

The Mandelbrot Set

In the last few chapters we have seen that the quadratic function $Q_c(z) = z^2 + c$ exhibits a vast number of different dynamical behaviors. Also, the Julia sets of Q_c vary greatly for different c-values. In this chapter we consider the question of how we understand these different shapes and dynamics. We will see that the Mandelbrot set provides a "dictionary" of all of these different structures. In a sense, the Mandelbrot set is a compilation of all of the different phenomena that we have already seen for the quadratic functions, and this set explains how all of these different structures and shapes are related.

We saw in the mathematical tour in Chapter 0 that the Mandelbrot set is one of the most intricate and beautiful objects in mathematics. We will see in this section that, despite its complexity, the Mandelbrot set is quite easy to compute. Moreover, each little bulb or antenna in the Mandelbrot set has a specific dynamical meaning that we can understand, given our previous work.

8.1   Critical Points and Orbits

To construct the Mandelbrot set, we need simply to understand the orbit of 0 under $Q_c(z) = z^2 + c$ for each different c-value. We have looked at this orbit before: Recall that when we plotted the orbit diagram of Q_c on the real line, it was the orbit of 0 that we plotted for each c. As we shall see, there is a reason why we chose 0 and not some other initial point. Since the orbit of

0 under Q_c plays a special or critical role in determining the dynamics, this orbit is called the *critical orbit*, and 0 is called the *critical point*.

For readers who know calculus, the precise reason why 0 is a critical point is the fact that the derivative of Q_c vanishes only at 0. For us, the reason why 0 is a special point stems from the fact that 0 is the only point that satisfies $Q_c(z) = c$; there is no other point z_0 in the complex plane for which

$$z_0^2 + c = c$$

Indeed, if we take any other point w_0 in the plane, there are always two points z_0 in the plane that satisfy $Q_c(z_0) = w_0$. We saw in the previous section that we can solve the equation

$$z_0^2 + c = w_0$$

by taking the two complex square roots of $w_0 - c$. These give two different values for z_0, provided $w_0 - c \neq 0$. Thus $w_0 = c$ is the only value in the complex plane that has just one preimage under Q_c, and that preimage is the critical point 0. We will see that the critical orbit determines virtually all of the dynamics of Q_c. That is, if we know what happens to the critical orbit, we can predict to a great extent the behavior of all other orbits.

We saw another instance of this in the previous section, where we saw that if the orbit of 0 tended to infinity, then the Julia set of Q_c was totally disconnected. We used the term fractal dust to describe the Julia set in this case. On the other hand, if the orbit of 0 did not escape, then the Julia set was connected — it consisted of exactly one piece. Another of the major reasons for the importance of the critical orbit is the following.

Fact. Suppose Q_c has an attracting periodic orbit. Then the critical orbit is attracted to this orbit.

Although we cannot prove this here, we note that this fact has the important consequence that Q_c can have at most one attracting periodic cycle, since the critical orbit can be attracted to at most one attracting cycle. We have already seen that some of these functions have infinitely many periodic orbits (recall that Q_{-2} has 2^n distinct periodic points that are fixed for Q_{-2}^n). So at most one of these orbits can be attracting, a somewhat surprising fact!

Given the importance of the orbit of 0 in determining the structure of the Julia set as well as in finding attracting cycles, it seems natural to ask which c-values have critical orbits that escape and which do not. As natural as this question seems, it is amazing that it was asked (and answered) only in 1978 by Benoit Mandelbrot. He was among the first to wonder what the set of c-values whose critical orbit does not escape looks like. Since he is a Research Fellow at IBM, he had the perfect tool nearby to look at this set, the computer. What he found has been called the most intricate object in all of mathematics; this set now bears his name, the Mandelbrot set.

8.2   Construction of the Mandelbrot Set

In this section, we describe the theory behind the computer program that produces the Mandelbrot set. To be precise, the *Mandelbrot set* is the set of c-values for which the critical orbit of Q_c does *not* tend to infinity. We call this set $\mathcal{M}$. As we will see, $\mathcal{M}$ is a dictionary that contains descriptions of all of the different dynamics that occur for the quadratic family. We emphasize the fact that the Mandelbrot set is a picture in the c-plane, not in the z-plane, where the Julia sets live.

What does $\mathcal{M}$ look like? To answer this we need to know which c-values have critical orbits that escape. It is easy to see that certain c-values lead to immediate escape of the critical orbit. For example, it is true that if $|c| > 2$, then the orbit of 0 escapes immediately. To see why this is true, we write $|c| = 2 + \ell$, with $\ell > 0$. We claim that *any* point z with $|z| \geq |c|$ escapes under iteration of Q_c (so, in particular, c itself escapes, and $c = Q_c(0)$, so the orbit of 0 escapes). This happens because if $|z| \geq |c| > 2$, then

$$
\begin{aligned}
|Q_c(z)| = |z^2 + c| &\geq |z|^2 - |c| \quad \text{by the triangle inequality} \\
&\geq |z|^2 - |z| \\
&= |z|(|z| - 1) \\
&\geq |z|(1 + \ell)
\end{aligned}
$$

This means that $|Q_c(z)| \geq |z|(1 + \ell)$ as long as $|z| \geq |c|$. But $1 + \ell > 1$. Therefore, $|Q_c(z)| > |z|$. This result may be interpreted as saying that as long as z lies outside the circle of radius $|c| > 2$ in the plane, then its image $Q_c(z)$ lies further away from the origin than z does. That is, under one iteration, such points move closer to infinity. Therefore, we may apply the same argument to say that under two iterations, such points move even further away from the origin. Precisely, we may apply the preceding reasoning to $Q_c^2(z)$ to find

$$
\begin{aligned}
|Q_c^2(z)| = |Q_c(Q_c(z))| &\geq |Q_c(z)|(1 + \ell) \quad \text{since } |Q_c(z)| \text{ is larger than 2} \\
&\geq |z|(1 + \ell)^2
\end{aligned}
$$

Here we have used the fact that $|Q_c(z)| > |z|(1 + \ell)$. Continuing in this fashion, we find

$$|Q_c^n(z)| \geq |z|(1 + \ell)^n$$

Now $1 + \ell > 1$. Therefore the nth power of $1 + \ell$ gets larger and larger as n increases. Indeed,

$$(1 + \ell)^n \to \infty$$

as $n \to \infty$. Therefore,

$$|Q_c^n(z)| \to \infty$$

as $n \to \infty$ too.

As a consequence, we now know that the Mandelbrot set lies within the circle of radius 2 in the complex plane. Equivalently, if $|c| > 2$, then the critical orbit of Q_c must escape, and the Julia set of Q_c is fractal dust. To find the Mandelbrot set, we therefore need only check c-values inside the circle of radius 2 to see if they belong to $\mathcal{M}$. This may be done by using our observation from Chapter 6: If the orbit of c ever leaves the disk of radius 2, then it necessarily escapes to infinity. This gives us the algorithm to compute the Mandelbrot set; we simply compute the orbit of 0:

$$0, c, \ c^2 + c, \ (c^2 + c)^2 + c, \ \left((c^2 + c)^2 + c\right)^2 + c, \ldots$$

and check whether any point on this orbit has modulus larger than 2. Once this occurs, we are guaranteed that the critical orbit escapes and c does not belong to the Mandelbrot set.

8.3 The Program MANDELBROT1

Let's now use these ideas to write a program called MANDELBROT1, which draws the Mandelbrot set. The algorithm is straightforward, given what we already know:

1. Divide the square $-2 \leq x, y \leq 2$ in the plane into a 300×300 grid.
2. Treat each grid point as a c-value.
3. For each such c, check whether the orbit of 0 under Q_c escapes within the first 30 iterations.
4. If so, color c white.
5. If not, color c black.

Figure 8.1 displays a BASIC program that uses this algorithm. Note the similarity between this program and our earlier JULIA1. The only difference here is subtle; for each grid point in the plane, we must remember the corresponding value of c throughout the iteration. That is why we use the statements

$$X = C1$$

$$Y = C2$$

at the outset and then perform all of the computations using x and y.

```
REM  program MANDELBROT1
CLS
FOR  i = 1 TO  300
   FOR  j = 1 TO  150
      c1 = -2 + 4 * i / 300
      c2 = 2 - 4 * j / 300
      x = c1
      y = c2
         FOR  n = 1 TO  30
         x1 = x * x - y * y + c1
         y1 = 2 * x * y + c2
         r = x1 * x1 + y1 * y1
         IF  r > 4 THEN GOTO  1000
         x = x1
         y = y1
         NEXT  n
      PSET (i, j)
      PSET (i, 300 - j)
1000 NEXT  j
   NEXT  i
END
```

Figure 8.1 The program MANDELBROT1.

We have again made use of a symmetry to speed up MANDELBROT1. Unlike the Julia sets of Q_c, $\mathcal{M}$ is not symmetric about the origin; $\mathcal{M}$ is not preserved by replacing both c_1 and c_2 by their negatives. However, the Mandelbrot set is symmetric about the x-axis in the following sense. Suppose we know that the point $c = c_1 + ic_2$ lies in the Mandelbrot set, with $c_2 > 0$. Then $c_1 - ic_2$ must also lie in $\mathcal{M}$.

To understand why this is so, we need to introduce the notion of the complex conjugate. If $z = x + iy$ is a complex number, its *complex conjugate* is a new complex number given by $\bar{z} = x - iy$. That is, $\bar{z}$ differs from z in that the sign of its imaginary part changes. Figure 8.2 shows the geometric interpretation of complex conjugation.

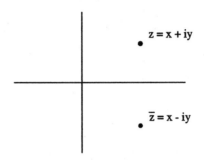

Figure 8.2 The complex conjugate of $z = x + iy$.

Complex conjugation has several simple algebraic properties that you may easily check. For example, if z and w are complex numbers, then

$$\bar{z} + \bar{w} = \overline{z + w}$$

$$\bar{z} \cdot \bar{w} = \overline{z \cdot w}$$

In particular, it follows that

$$
\begin{aligned}
\overline{Q_c(z)} &= \overline{z^2 + c} \\
&= \overline{(z^2)} + \bar{c} \\
&= \overline{z} \cdot \overline{z} + \bar{c} \\
&= (\bar{z})^2 + \bar{c} \\
&= Q_{\bar{c}}(\bar{z})
\end{aligned}
$$

That is, the complex conjugate of the image of z under Q_c is precisely the same as the image of $\bar{z}$ under $Q_{\bar{c}}$. This simply means that if we know the orbit of z under Q_c, we can obtain the orbit of $\bar{z}$ under $Q_{\bar{c}}$ by simply taking the complex conjugate at each stage. See Figure 8.3. In particular, the orbit

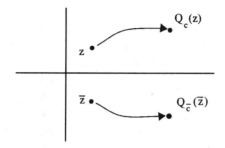

Figure 8.3 The orbits of z under Q_c and $\bar{z}$ under $Q_{\bar{c}}$.

of $\bar{c}$ under $Q_{\bar{c}}$ is precisely the complex conjugate of the orbit of c under Q_c. Thus if the critical orbit of Q_c escapes, so too does the critical orbit of $Q_{\bar{c}}$.

We have used this fact to cut in half the number of computations necessary to produce $\mathcal{M}$. Indeed, if the pixel corresponding to c is colored black, then we immediately color $\bar{c}$ black too. This is accomplished by the statements

$$\text{PSET}(i, j)$$
$$\text{PSET}(i, 300 - j)$$

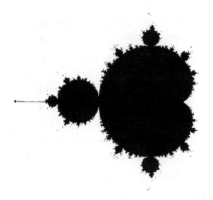

Figure 8.4 The Mandelbrot set.

Figure 8.4 displays the image generated by MANDELBROT1. Remember that this is a picture in the c-plane — the parameter plane. Unlike the situation for Julia sets, where we get a different picture for each different c-value, here there is only one image. In $\mathcal{M}$, each pixel that is colored black corresponds to a quadratic function whose Julia set is connected. As we have seen, these Julia sets come in a variety of shapes and sizes. The Mandelbrot set tells us which c-values correspond to Julia sets with the same basic shape and which quadratic functions have more or less the same dynamics.

Note that $\mathcal{M}$ consists of a basic cardioid-shaped region with many "decorations," or bulbs, attached. The cusp of the cardioid occurs precisely at the point $c = \frac{1}{4}$, which is the value that corresponds to the saddle-node bifurcation. Directly to the left of the main cardioid is a large, circular region. This bulb is attached to the cardioid at $c = -\frac{3}{4}$, which is the period–doubling bifurcation point. Finally, there appears to be a tail, or antenna, which emanates from $\mathcal{M}$ and points toward the left. This tail lies along the real axis and terminates at exactly $c = -2$, which is the parameter value for which we found infinitely many periodic points earlier. We will return to make a much more detailed study of $\mathcal{M}$ and what all of these decorations and antennae mean, but first let's modify MANDELBROT1 so that we can see the fine detail in $\mathcal{M}$.

8.4 Refinements of MANDELBROT1

We will often return to the Mandelbrot set in the remainder of this chapter. As you may have noticed, it takes quite a while to compute the image of $\mathcal{M}$. Therefore, it would be wise to save the image in a file rather than recompute it each time we need to look at it.

There are a number of ways to save the image. One way that is highly inefficient is to save the number of iterations necessary to "fail" the test in MANDELBROT1. That is, for each pixel tested, we record the integer value at which our program stops computing the orbit of c. This will result in a 300×300 array of integers. Given 2 bytes of storage per integer, this means we will have to store 1.8×10^5 bytes of information. While this is not a prohibitive amount of information to be stored using today's technology, we can compress this data significantly.

Project 8.1 Devise an efficient method for storing your image of the Mandelbrot set so that you can redisplay it on the screen at will, without recomputing all of the orbits.

Saving the iteration count (as opposed to a simple black/white indicator) will be useful if you use some of the later programs that utilize color.

As in the case of Julia sets, there are more sophisticated algorithms that produce sharper images of the Mandelbrot set. We discuss some of them next. Our goal is to understand what the Mandelbrot set means. To comprehend the relationship between Julia sets and the Mandelbrot set, it is helpful to be able to magnify certain portions of $\mathcal{M}$. This is accomplished by the following program.

Project 8.2 Write a program called MANDELBROT2 that allows the user to select a square from the output of MANDELBROT1. The new program then recomputes the portion of $\mathcal{M}$ in a 300×300 grid inside this square. Your program should allow the user to select and compute in successively smaller and smaller squares in order to visualize the fine structure of $\mathcal{M}$. Since many more than 30 iterations will be necessary to view $\mathcal{M}$ in fine detail, your program should allow the user to select the maximum number of points on the critical orbit that will be computed before coloring a pixel white.

Remember that the symmetry exploited in MANDELBROT1 will no longer be valid for this program (unless the chosen square is symmetric about the x-axis). So you will have to compute the critical orbits for all of the c-values in your square and eliminate the second PSET command.

Experiment 8.3 Use MANDELBROT2 to compute the portion of $\mathcal{M}$ that lies inside the following small squares in the c-plane.
 a. The box with side of length .02 centered at $-1.256 + .38i$.
 b. The box with side of length .03 centered at $-1.185 + 0.3i$.

Outcome. In both cases, you will see small copies of the Mandelbrot set inside these windows. See Figure 8.5. This is one of the truly remarkable and surprising features of $\mathcal{M}$. All the complexity that occurs in the "main body" of $\mathcal{M}$ occurs as well within these smaller copies. So there are baby Mandelbrot sets within these small copies of Mandelbrot set, and so on. It is difficult to see these smaller Mandelbrot sets without resorting to double–precision computations and higher–resolution screens, however.

There are many, many other interesting regimes in the Mandelbrot set that you can explore with this program.

Experiment 8.4 Use MANDELBROT2 to compute various portions of the $\mathcal{M}$. Keep a record of the coordinates of each image you generate as well as

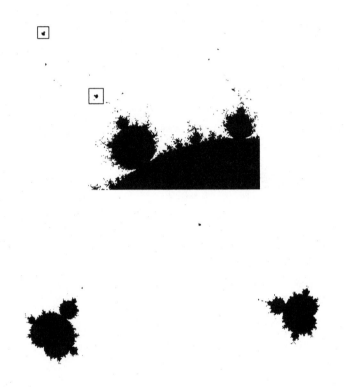

Figure 8.5 Baby Mandelbrot sets within $\mathcal{M}$.

a picture book of these images. Can you find the various regimes depicted in the color plates in Chapter 0?

Outcome. Two portions of $\mathcal{M}$ are displayed in Figure 8.6. The first image here is the small bulb that sits on top of the main cardioid in $\mathcal{M}$. This is called the period 3 bulb, for reasons that we will explain later. The coordinates in the complex plane are $-.4 \le x \le .2$ and $.6 \le y \le 1.2$. The second image is that of one of the bulbs located near the cusp in the main cardioid. The coordinates are $.385 \le x \le .485$ and $.17 \le y \le .27$.

Figure 8.6 Detail of the Mandelbrot set.

Despite appearances to the contrary, the Mandelbrot set is a connected set. This is a remarkable recent theorem, which was proved by the mathematicians Adrien Douady and John Hubbard in 1982. Note that Figures 8.4 and 8.5 seem to suggest that $\mathcal{M}$ has small "islands," which are not attached to the main body. In fact, however, these islands are connected to the main body by thin filaments that are invisible at this resolution.

The boundary scanning method used to display Julia sets may also be used to generate the Mandelbrot set. This method makes use of the following algorithm:

1. Divide the square $-2 \leq x, y \leq 2$ into a 300×300 grid in the plane. For each pixel (m, n) in this grid:

2. Let c be the value in the plane corresponding to (m, n).

3. Compute the first 20 points on the critical orbit of Q_c.

4. If any point on this orbit has modulus larger than 2, then color (m, n) white.

5. If this orbit does not escape, then compute the critical orbits of the c-values corresponding to the four pixels $(m+1, n)$, $(m-1, n)$, $(m, n+1)$, and $(m, n-1)$.

   a. If at least one of these additional orbits escapes, then color (m, n) black.

   b. If all four of these additional orbits do not escape, then color (m, n) white.

This method yields a quite different picture of $\mathcal{M}$; only the boundary of $\mathcal{M}$ is colored. Can you explain why?

Project 8.5 Use the boundary–scanning method to write a program called MANDELBROT3. Use this program to explore various regions of $\mathcal{M}$. It is quite interesting to examine the bulbs of $\mathcal{M}$ using this method, especially those near the cusp of the main cardioid.

The image of the Mandelbrot set generated by MANDELBROT3 is displayed in Figure 8.7.

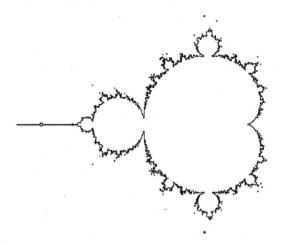

Figure 8.7 $\mathcal{M}$ computed by the boundary-scanning method.

8.5   What the Mandelbrot Set Means

Now let's write the granddaddy of all Mandelbrot set programs. This program will allow us to understand the relationship between the Julia sets for Q_c and the corresponding c-values in $\mathcal{M}$.

Project 8.6 Construct a program called MANDELBROT4, which builds on MANDELBROT2 and allows the user to select a particular c-value from the image displayed by MANDELBROT2. The user should be able to select this c-value using a mouse or cross hairs, if possible, or else by using the keyboard to choose particular screen coordinates. In any event, once a particular c-value has been selected, the user should be given the choice of selecting one of our previous programs, ITERATE5 or JULIA2, as applied to Q_c for the chosen c-value. The screen should then be cleared and the appropriate orbit or Julia set displayed. Note that it is crucial that we be able to restore M to the screen quickly in order to be able to use this program effectively; recomputing M each time wastes a lot of time!

Experiment 8.7 Use MANDELBROT4 to compute the orbit of c and the Julia set of Q_c, where c is any value in the cardioid-shaped region in M (the largest black bulb with the cusp at $c = \frac{1}{4}$).

Outcome. For each c-value in this region, the critical orbit is attracted to an attracting fixed point and the Julia set is a closed curve.

Experiment 8.8 Use MANDELBROT4 to compute the orbit of c and the Julia set of Q_c, where c is any value in the large, circular bulb immediately to the left of the cardioid.

Outcome. The critical orbit is attracted to an attracting cycle of period 2 and the Julia sets resemble the Julia set of $Q_{-1}(z) = z^2 - 1$.

Experiment 8.9 Use MANDELBROT4 to compute the critical orbit of Q_c when c is inside one of the bulbs in M.

Outcome. You will see that, as long as c is strictly within one of the bulbs, the critical orbit is attracted to an attracting cycle. The period of this cycle depends very much upon which bulb c lies in, but it remains the same for all c-values in a given bulb. For example, in the small ball directly to the left of the period two regime discussed in the last experiment, the critical orbit is attracted to a cycle of prime period 4.

Experiment 8.10 What happens to the critical orbit when c is inside one of the two largest balls attached to the top and bottom of the main cardioid?

If you have stored other images of portions of M such as those in Figure 8.5 or 8.6, you can further investigate the relationship between M and the Julia sets. For example, what happens when c lies inside the main cardioids of each of the baby Mandelbrot sets in Figure 8.5?

Thus we see that the Mandelbrot set contains dynamical information about the behavior of the critical orbits and the shape of the Julia set. Inside each of the black bulbs in M, the critical orbit behaves in the same manner. Usually, this means that the critical orbit is attracted to a cycle of some fixed period. To be honest, nobody knows whether this *always* happens, but extensive numerical experimentation suggests that this is true.

Thus we may assign to each bulb in $\mathcal{M}$ a number that corresponds to the period of the attracting cycle for each Q_c with c in that bulb.

Experiment 8.11 Compute the period of the attracting cycle for as many bulbs in $\mathcal{M}$ that your computer allows you to see.

Outcome. Partial results of this experiment are displayed in Figure 8.8.

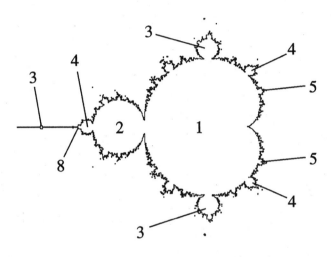

Figure 8.8 The periods of the bulbs in $\mathcal{M}$.

Recall the orbit diagram for the quadratic family $Q_c(x) = x^2 + c$ that we constructed in Chapter 4. To construct this diagram, we looked at the last 150 of 200 points on the orbit of 0 under iteration of Q_c for a real value of c. This is essentially what we did to compute $\mathcal{M}$. This means that there is a direct relationship between the orbit diagram and the horizontal slice through the middle of $\mathcal{M}$. After all, this horizontal line tells us what happens to the critical orbit for real values of c.

Figure 8.9 juxtaposes these two images. The real c-values for the orbit diagram and $\mathcal{M}$ are each plotted horizontally; the scale has been chosen so that c increases from -2 to .5 in each picture with corresponding c-values located directly above and below one another. Note that the attracting fixed point regime corresponds exactly to the main cardioid, as we know from our preceding experiment . Similarly, the period 2 and 4 regimes correspond. Note that the period 3 window in the orbit diagram sits directly above the baby Mandelbrot set located in the tail of $\mathcal{M}$. This explains why this period 3

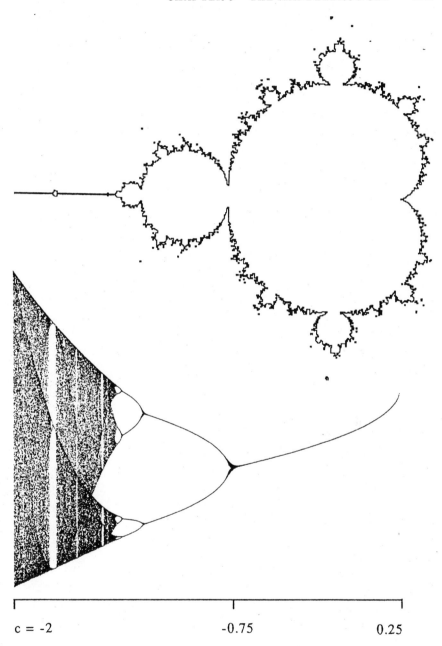

$$c = -2 \qquad\qquad -0.75 \qquad\qquad 0.25$$

Figure 8.9 The orbit diagram and $\mathcal{M}$.

cycle undergoes period doubling as c decreases; the parameter passes through all the period–doubling bulbs of this baby Mandelbrot set as c decreases.

There is much more to this story, as is obvious from the orbit diagram. Many of the critical orbits do not seem to find an attracting cycle. This is true no matter how many iterations you choose in the program ORBITDGM. These c-values do not lie in the bulbs attached to $\mathcal{M}$. Rather, they lie on the thin filaments and correspond to c-values for which there are no attracting cycles whatsoever. These c-values yield dynamics that are extremely complicated. They are the parameter values for which Q_c is chaotic. We will spend most of Chapter 10 investigating what this term means. For now we simply note that, as is apparent from Figure 8.9, many of the c-values between -2 and $.25$ lead to chaotic dynamics.

Further Exercises and Experiments

1. Write a program similar to MANDELBROT1 to compute the analogue of the Mandelbrot set for the function $F_c(z) = \overline{z}^2 + c$, where $\overline{z}$ is the complex conjugate of z as introduced in Section 8.3. This set is called the "tricorn" because of its resemblance to a tricornered hat.

2. If you have lots of friends with computers, you can make a large "mural" of the Mandelbrot set by slicing up the region $|c| \leq 2$ into many squares and then computing the portion of $\mathcal{M}$ inside each square. You can then cut and paste together all the output from this experiment to obtain a detailed picture of the Mandelbrot set. Of course, your friends who are assigned all-black regions of $\mathcal{M}$ will not have too much fun.

3. Compute the analogue of the Mandelbrot set for the complex logistic function $F(z) = cz(1 - z)$. In this case, the critical point is $\frac{1}{2}$, not 0, so you must compute the orbit of $\frac{1}{2}$ to see if it escapes. One test to decide if orbits of the logistic map escape was given in Problem 2 of Chapter 6.

Chapter 9

Geometric Iteration: Fractals

The word *fractal* has arisen several times thus far in our discussion of dynamical systems. We used fractal to describe Julia sets which, when magnified over and over again, always seemed to look the same. We also encountered this term when Julia sets disintegrated into fractal dust. In this chapter we delve more deeply into the amazingly complicated yet regular structure of these sets. We show how to create a large variety of fractals and we discuss how these sets naturally have "fractional" dimension.

9.1   Fractals

There are many mathematical processes that may be iterated. We have already seen how iterated functions give rise to complicated dynamics. Now we will iterate certain geometric constructions. These iterations will yield the complicated geometric objects known as fractals.

The subject of fractal geometry was given its name by Benoit Mandelbrot in the mid–1970s. Actually, this subject has a long and interesting history in mathematics, involving many mathematicians from many parts of the world over the last centuries, including Peano from Italy, Cantor and Hausdorff from Germany, Besicovitch from Russia, and many others. Mandelbrot realized that the bizarre, seemingly contrived geometric constructions engineered by these mathematicians were not at all pathological, as they were at first regarded. Rather, he showed that many everyday objects such as coastlines, snowflakes, clouds, leaves, ferns, and mountain ranges were naturally described by fractals. Ordinary geometric constructions using straight lines and smooth curves and surfaces did not help very much to understand or to model many of the intricate patterns found in nature. Thus was born a new branch of mathematics, fractal geometry. In its short lifespan, this field of mathematics has proved to be extremely useful in a variety of applications. Even Hollywood has made use of fractal geometry to create fractal landscapes and planetscapes for films.

What is a fractal? Basically, a fractal is a geometric shape that has two special properties:

1. The object is *self-similar*.
2. The object has fractional dimension.

In the next few sections we present several geometric constructions of fractals to illustrate the concept of self-similarity. Then we move on to the more difficult question of how an object can have fractional dimension.

9.2   The Sierpinski Triangle

The Sierpinski triangle is generated by an infinite succession of "removals." We begin with an equilateral triangle. We then remove the middle upside-down triangle, as depicted in Figure 9.1. This leaves three smaller equilateral triangles untouched. From each of these triangles we then remove the middle equilateral triangle, which then leaves nine smaller equilateral triangles. We then iterate this procedure: from each remaining equilateral triangle we remove the middle triangle, leaving three smaller triangles behind.

Exercise 9.1 How many triangles remain after the third collection of triangles is removed? After the fourth? After the fifth? Can you find a formula for the number of triangles which remain after the n th stage of this process?

The ultimate figure that results from all of these removals is called the Sierpinski triangle. Notice that this "triangle" has no two-dimensional regions whatsoever — any planar piece of the original triangle is eventually decimated by the removal of small triangles. See Figure 9.2.

Why is this figure self-similar? To understand this, let's look closely at the lower left portion of the triangle, the portion contained within one of the three triangles that remain after the first step of the process. Notice that the length of a side of this triangle is exactly one–half the length of a side of the original triangle. Notice also that we have removed infinitely many triangles from this smaller triangle, just as we have from the original one. Moreover, if

Figure 9.1 Construction of the Sierpinski Triangle.

Figure 9.2 The Sierpinski triangle.

we peer at this triangle through a microscope that magnifies all dimensions by a factor of two, then what we see is precisely the entire Sierpinski triangle! That is, this small, lower left portion of the triangle is exactly the same as the whole triangle when magnified by a factor of two! This is by no means the end of the story. Suppose we take any of the nine triangles that remain after the second stage of removals. If we magnify this triangle by a factor

of four, we again find the original triangle, and if we magnify the portion of
the Sierpinski triangle inside this small triangle by four, we again find the
original figure.

This continues no matter how "deep" we look within the triangle. The
portion of the triangle contained within a triangle at level n, when magnified
by a factor of 2^n, is exactly the same as the whole triangle. This is self-
similarity — small portions of the triangle, when magnified, are similar to
the whole triangle.

Exercise 9.2 Carefully draw the next stages of the figure that is obtained
by removing four smaller squares from a square, as depicted in Figure 9.3.
How many squares are left after the third removal? After the fourth? The
result of this operation is depicted in Figure 9.4.

Figure 9.3 Construction of the box-fractal.

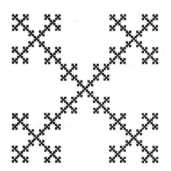

Figure 9.4 The box-fractal.

9.3   The Cantor Set

In many senses, the simplest of all fractals is the Cantor set. It is also one whose "cousin" we encountered when we found Julia sets that were fractal dust. To construct this set we begin with the interval $0 \leq x \leq 1$. From this we remove the middle third: the interval $\frac{1}{3} < x < \frac{2}{3}$. Note that we leave the endpoints behind. What remains is a pair of intervals, each of them one–third as long as the original. Now we do this again: From each of the remaining intervals we remove the middle third. That is, from the interval $0 \leq x \leq \frac{1}{3}$, we remove $\frac{1}{9} < x < \frac{2}{9}$, and from $\frac{2}{3} \leq x \leq 1$ we remove $\frac{7}{9} < x < \frac{8}{9}$. Note that, in each case, we have again left behind the endpoints; now four intervals remain, each one–ninth as long as the original interval.

We now repeat this process over and over. At each stage, we remove the middle third of each of the intervals remaining at the previous stage. What is left when we are finished is the Cantor middle-thirds set, or, for short, the Cantor set. See Figure 9.5.

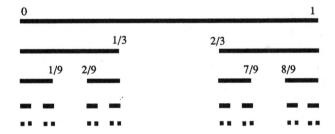

Figure 9.5 Construction of the Cantor middle-thirds set.

Exercise 9.3 Identify the endpoints of the intervals that are removed at the third stage of the process. How many intervals remain behind? Identify the endpoints of the intervals that are removed at the next two stages of this process.

At each stage, the number of intervals that are left behind increases, but they become small quite quickly. There are $2 = 2^1$ intervals of length $\frac{1}{3}$ after the first removal, 2^2 intervals of length $\frac{1}{3^2}$ after the second, 2^3 of length $\frac{1}{3^3}$

after the third, and so forth. For example, there are 1024 intervals after the tenth stage of this process, but each has length

$$\frac{1}{3^{10}} = \frac{1}{59,049} = .0000169\ldots$$

Note that there are points that are never removed during this process, so the Cantor set is not an empty set. For example, any endpoint of a removed interval belongs to the Cantor set. This is true, since at each stage we remove an interval that is close to, but never contains, this endpoint. So it follows that $0, 1, \frac{1}{3}, \frac{2}{3}, \frac{1}{9}, \frac{2}{9}, \frac{7}{9}$, and $\frac{8}{9}$ are all points in the Cantor set.

Note also that the Cantor set contains no intervals, for if it did, at some stage we would have removed its middle third. This means, to use a word we introduced in Chapter 7, that the Cantor set is totally disconnected: Between any two points in the set, there must be points that do not belong to the set.

The Cantor set is also self-similar. To see this, look closely at the left interval $0 \leq x \leq \frac{1}{3}$, which remains behind after the first interval is removed. There is a portion of the Cantor set contained within this interval. If we examine this portion of the Cantor set using a microscope that magnifies by a factor of three, then what we see is an exact replica of the full Cantor set!

The same thing is true if we magnify the right–hand interval by a factor of three. If we zoom in on any of the intervals remaining at the second stage, for example $0 \leq x \leq \frac{1}{9}$, then magnification by a factor of 9 reveals the full Cantor set again. See Figure 9.6. You should compare this with Figure 7.6.

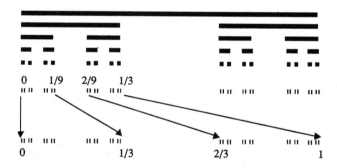

Figure 9.6 Magnification of the Cantor set by a factor of 3.

The action of viewing a Cantor set under a microscope can be made mathematically precise, for it corresponds exactly to multiplication by a power of 3. For example, what happens if we multiply each number in $0 \le x \le \frac{1}{3}$ by 3? The operation of multiplying by 3 stretches this small interval over the entire interval $0 \le x \le 1$. Moreover the middle third, $\frac{1}{9} < x < \frac{2}{9}$ of the small interval corresponds exactly to the middle third, $\frac{1}{3} < x < \frac{2}{3}$, of the larger interval. Subsequent middle thirds correspond as well. So we see that magnification of this interval is really just multiplication by 3.

It is interesting to ponder this fact for a while. It may seem that there are exactly one half the number of points in the Cantor set that lie in the left–hand interval. But magnification shows that there are, in fact, just as many points in this side of the Cantor set as there are in the whole set — this small portion of the Cantor set is exactly the same as the whole set!

Exercise 9.4 You can construct another Cantor set by removing middle fifths of intervals, instead of middle thirds. That is, we remove the interval $\frac{2}{5} < x < \frac{3}{5}$ from $0 \le x \le 1$ at the first stage, then remove the middle fifth of each remaining interval, and so forth. List the intervals that are removed in the first few stages of this construction.

It may appear that the only points in the Cantor set are endpoints of the intervals that have been removed. Actually, this is far from the truth: There are many, many other points in the Cantor set that are not endpoints. In fact, *most* points in the Cantor set are not endpoints! Here is how to see this. Each time we remove an interval, we leave behind two other intervals, one on the left and one on the right. If a point lies in the Cantor set, then we may identify it by giving the sequence of lefts and rights that indicate to which interval the point belongs as each successively smaller middle interval is removed. For example, 0 corresponds to all lefts, because 0 is always in the left–hand subinterval as each middle third is removed. So we write

$$0 \mapsto LLLL\ldots$$

to indicate this. Similarly, 1 is always in the right hand interval, so

$$1 \mapsto RRRR\ldots$$

You may easily check that

$$\frac{1}{3} \mapsto LRRR\ldots$$

because $\frac{1}{3}$ lies in the original left interval, but thereafter it is always the rightmost endpoint of any interval which undergoes a removal. Also,

$$\frac{2}{3} \mapsto RLLL\ldots$$

$$\frac{1}{9} \mapsto LLRRR\ldots$$

$$\frac{2}{9} \mapsto LRLLL\ldots$$

Exercise 9.5 Identify the sequence of lefts and rights corresponding to $\frac{7}{9}, \frac{8}{9}, \frac{1}{27}, \frac{7}{27}, \frac{25}{27}, \frac{19}{81}$ and $\frac{20}{81}$.

Notice one fact. Each endpoint of an interval corresponds to a sequence that consists of a finite number of lefts or rights, followed by an infinite string that is either all lefts or all rights. This means that any point corresponding to a sequence that does not terminate in all lefts or all rights cannot possibly be an endpoint. Do you see why?

You are right to ask whether or not there are any such points. Indeed there are! Here's how to see that there is a point that corresponds to the repeating sequence $LRLRLR\ldots$. Start with the left interval $0 \leq x \leq \frac{1}{3}$. This interval contains the right–hand subinterval $\frac{2}{9} \leq x \leq \frac{3}{9}$ constructed at stage 2. This interval, in turn, contains the left–hand interval $\frac{6}{27} \leq x \leq \frac{7}{27}$ constructed at stage 3. The right–hand interval constructed at stage four is $\frac{20}{81} \leq x \leq \frac{21}{81}$. Can you continue this process? Do you see the pattern?

Exercise 9.6 Identify the intervals corresponding to the repeating sequence $LRLRLR\ldots$ at the fifth, sixth, and seventh stages of this process.

Note that at the k th stage of this construction, we find an interval of length $\frac{1}{3}^k$ that is contained inside the previous interval. These intervals decrease in size until, when the process is completed, only one point remains. This is the point that corresponds to the sequence $LRLRLR\ldots$. Note that this point can't be an endpoint of a removed interval because the sequence $LRLRLR\ldots$ does not end in a constant sequence.

Remark. Amazingly, you can compute the point that corresponds to this sequence exactly! The number with sequence $LRLRLR\ldots$ is $\frac{1}{4}$. Similarly, $\frac{3}{4}$ corresponds to the sequence $RLRLRL\ldots$. To understand this, you need to know how to compute infinite series, a topic usually covered in calculus.

Similar arguments show that there is a point in the Cantor set that corresponds to *any* sequence of lefts and rights. Clearly, there are many, many sequences that do not correspond to endpoints; there is more to a Cantor set than first meets the eye!

9.4 The Koch Snowflake

Unlike the Sierpinski triangle, the Koch snowflake is generated by an infinite succession of additions. This time we begin with the boundary of an equilateral triangle with side of length 1. The first step in the process is to remove the middle third of each side of the triangle, just as we did in the construction of the Cantor set. This time, however, we replace each of these pieces with two pieces of equal length, giving the star-shaped region depicted in Figure 9.7. This new figure has 12 sides, each of length $\frac{1}{3}$. Now we iterate this process. From each of these 12 sides we remove the middle third and replace it with a triangular "bulge" comprised of two pieces of length 1/9. The result is also shown in Figure 9.7.

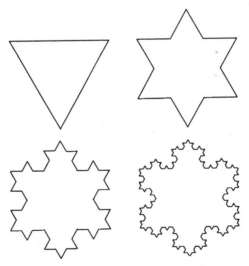

Figure 9.7 The first four stages
in the construction of the Koch snowflake.

We continue this process over and over. The ultimate result is a curve that is infinitely wiggly — there are no straight lines in it whatsoever. This object is called the Koch snowflake.

Why is this snowflake self-similar? Suppose we look at one side of the original triangle. What we see is depicted in Figure 9.8. If we examine one third of this edge and magnify this portion by a factor of three, we again see the same figure.

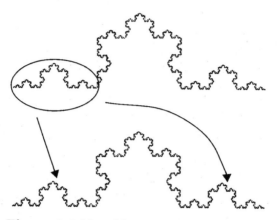

Figure 9.8 Magnification of the Koch curve.

At each stage of the construction of the Koch curve, magnification by a factor of 3 yields the previous image. As before, this means that the ultimate figure is self-similar.

This fractal has an amazing geometric property: It has finite area but its perimeter is infinite! This means that we can paint the inside of the Koch snowflake, but we can never wrap a length of string around its boundary! This is quite a contrast to the usual shapes encountered in geometry such as squares and circles, which have finite area *and* perimeter.

Let's see why this is true. Let $N_0, N_1, N_2, \ldots$ denote the number of sides of the snowflake at the corresponding stage of the construction. We find

$$N_0 = 3$$
$$N_1 = 4 \cdot 3 = 12$$
$$N_2 = 4 \cdot 12 = 4^2 \cdot 3$$
$$N_3 = 4^3 \cdot 3$$

We obtain the number at the kth stage by simply multiplying N_{k-1} by 4, since our procedure calls for subdividing each side into 4 smaller ones.

Therefore we have

$$N_k = 4N_{k-1} = 4^k \cdot 3$$

These numbers get large very quickly. For example, N_9 gives 196,608 little sides!

Now let's compute the perimeter. Let L_k be the length of one segment of the perimeter after the kth stage. Notice that in the beginning, each side has length 1; after the first addition, each side has length $\frac{1}{3}$; after the second, each side has length $\frac{1}{3}^2$, and so forth. We find that, at the kth stage,

$$L_k = \frac{1}{3^k}$$

Now let P_k be the perimeter of the figure at the kth stage. Clearly, $P_k = N_k \cdot L_k$. That is, P_k is simply the product of the number of segments at the kth stage and the length of each of these segments. We find

$$P_k = N_k \cdot L_k = 4^k \cdot 3 \cdot \frac{1}{3^k}$$
$$= \left(\frac{4}{3}\right)^k \cdot 3$$

Note that, from this formula, each P_k is larger than its predecessor by a factor of $\frac{4}{3}$. So we see that the perimeter of the snowflake keeps getting larger and larger by a factor of $\frac{4}{3}$ as the number of stages grows. Hence, the ultimate perimeter must be infinite!

The area contained within the snowflake is more difficult to compute, but you can easily check using plane geometry that the snowflake is contained within a square in the plane whose sides have length $2\sqrt{3}/3$. Do you see why? Therefore this area is less than $\frac{4}{3}$.

Project 9.7 Can you think of other fractals that are generated by additions like the Koch snowflake? Draw the first few steps in the construction.

9.5   Computing Fractals

Certain fractals such as the Cantor set or the Sierpinski triangle are easy to draw using a computer and an iterative procedure similar to the backward iteration method of Chapter 7. We will obtain these fractals as the orbit of a single point under the iteration of a collection of functions. One major difference between this process and our previous work is that these functions will be iterated in random order. A second difference is that we will define these functions directly in screen coordinates rather than on the line or plane. This will eliminate the need to change coordinates in order to plot the fractals.

Let's begin with a simple example. Suppose we consider the three points $A = (0,0)$, $B = (0,300)$, and $C = (300,300)$ on the screen. We will define three functions F_A, F_B, and F_C. In words, F_A is given by the rule: Take any point (M,N) on the screen and move it to the point halfway between A and (M,N). F_B and F_C are defined analogously.

What are the formulas for these functions? As we see in Figure 9.9, the midpoint between A and (M,N) is simply $(M/2, N/2)$. Of course, one or both of $M/2$ and $N/2$ may not be an integer. This will not matter to us because, when we plot these points via the $PSET$ command, they will be rounded to the nearest integer.

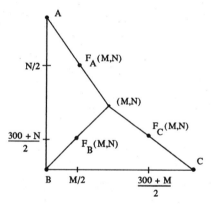

Figure 9.9 The functions F_A, F_B, and F_C.

The midpoint between B and (M,N) is given by

$$F_B(M,N) = (\frac{M}{2}, \frac{300+N}{2})$$

The midpoint between C and (M,N) is

$$F_C(M,N) = (\frac{300+M}{2}, \frac{300+N}{2})$$

If we draw the right triangle T with vertices A, B, and C, then the geometric interpretation of these functions is depicted in Figure 9.10. Each function compresses T into a new triangle exactly half the size of T and

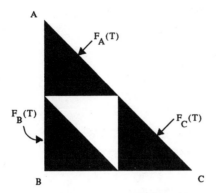

Figure 9.10 The image of T under F_A, F_B, and F_C.

having exactly one vertex in common with T. For example, $F_A(T)$ is the small right triangle with one vertex at A, as shown in Figure 9.10.

Now let's iterate these three functions in random order. More precisely, we will input any point (M, N) in screen coordinates. Then we will randomly select a number q between 0 and 3. If $0 \le q < 1$, then we will apply F_A; if $1 \le q < 2$, we will apply F_B; and if $2 \le q < 3$, we will apply F_C. The resulting iteration gives us what is known as a *random orbit* of (M, N). As before, in plotting this orbit, we will not plot the first few points on the random orbit.

In Figure 9.11 we have listed a program called FRACTAL, which implements this algorithm. Several remarks are in order. In this program, the function RND returns a random number between 0 and 1. Hence $3 * RND$ returns a random number q between 0 and 3. Depending on the value of q, we branch to compute the function F_A, F_B, or F_C. Instead of employing this time-consuming branching process, we could have replaced the branches by the single statements

$$q = INT(3 * RND)$$
$$M = M/2 + q * (q - 1) * 75$$
$$N = N/2 + q * (3 - q) * 75$$

Note that in this formulation, q is an integer, either 0, 1, or 2. The new value of M is $M/2$ if $q = 0$ or $q = 1$, but if $q = 2$, then the new value of M is

$$\frac{M}{2} + 150 = \frac{M + 300}{2}$$

Similarly, the new value of N assumes the correct form if $q = 0, 1,$ or 2.

```
REM program FRACTAL
INPUT  "M"; M
INPUT  "N"; N
CLS
FOR I = 1 TO 100000
      q = 3 * RND
      IF q < 1 GOTO 10
      IF q < 2 GOTO 20
      M = (300 + M) / 2
      N = (300 + N) / 2
      GOTO 100
10    M = M / 2
      N = N / 2
      GOTO 100
20    M = M / 2
      N = (300 + N) / 2
100 IF I < 1000 THEN GOTO 200
PSET (M, N)
200 NEXT i
END
```

Figure 9.11 The program FRACTAL.

Experiment 9.8 Use FRACTAL to compute a random orbit of any point (M, N). What do you observe for different choices of M and N?

Outcome. No matter which M and N are input, the same figure results. See Figure 9.12. Note that this image is a fractal and bears a strong resemblance to the Sierpinski triangle.

It is quite amazing that the picture resulting from the random iteration in FRACTAL does not seem to depend on M or N. Virtually any random orbit yields the same figure, at least at the resolution of the screen. You can define a similar set of functions in the plane: In this case the random orbit would lie in the plane and assume noninteger values. But again, virtually the same picture will result from any random orbit.

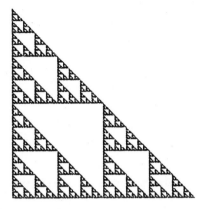

Figure 9.12 The output of FRACTAL.

The collection of functions that we have iterated randomly is called an *iterated function system*. Recent research of Barnsley and others has shown that these kinds of dynamical systems are quite important in such applications as image and data compression. The reason for this is that a complicated picture like Figure 9.12 can be stored in memory using only the defining function F_A, F_B, and F_C rather than storing the coordinates of all the points that make up the picture.

The fractal that results from this process is called an *attractor* because it attracts all of the random orbits. That is the reason why we drop the first 1000 points on a random orbit: We wish to see only the attractor.

Clearly, it is easy to modify FRACTAL to obtain a variety of interesting images.

Project 9.9 Modify FRACTAL so that it randomly iterates one of the following five functions. Let $A = (0,0)$, $B = (300,0)$, $C = (300,300)$, $D = (0,300)$, and $E = (150,150)$. Each of the functions F_A through F_E moves a point (M, N) two-thirds of the way toward the corresponding point A–E. For example,

$$F_E(M,N) = ((2*150 + M)/3, (2*150 + N)/3)$$
$$F_D(M,N) = (M/3, (2*300 + N)/3)$$

What fractal results when you run this program? What happens if you leave out one of the functions such as F_A?

Exercise 9.10 Find an iterated function system that produces the Sierpinski triangle as its output. Find another system that produces the middle-thirds Cantor set.

Exercise 9.11 Use the methods of this section to construct a variety of fractals. Once you select the iterated function system, can you predict ahead of time what the resulting attractor will look like?

9.6   Fractional Dimension

How do we assign a dimension to a geometric object? For certain familiar figures like lines or squares or cubes, it's easy. We almost intuitively feel that a line has dimension one, a square dimension two, and a cube dimension three. Perhaps that's because we feel that there is essentially only one direction that we can move along on a line, two directions in a square, and three in a cube. This is fine, but how do we use this idea to calculate the dimension of the Sierpinski triangle? At times, we feel that we can move in lots of planar directions on this triangle; at other times it seems like the final shape is only one-dimensional. So what *is* the dimension? Actually, it is somewhere in between, just as our eye and our "feelings" are telling us.

To see why this should be true, let's investigate the notion of dimension of lines, squares, and cubes more thoroughly. One way to realize that these objects have different dimensions is to do the following. A line is a very self-similar object: It may be decomposed into $n = n^1$ little "bite-size" pieces, each of which is exactly $\frac{1}{n}$ the size of the original line and each of which, when magnified by a factor of n, looks exactly like the whole line. On the other hand, if we decompose a square into pieces that are $\frac{1}{n}$ the size of the original square, then we find we need n^2 such pieces to reassemble the square. Similarly, a cube may be decomposed into n^3 pieces, each $\frac{1}{n}$ the size of the original. See Figure 9.13. So, here is one way to distinguish the dimension of an object: The exponent in each of these cases is precisely the dimension.

In these simple cases, it is trivial to read the exponent and find the dimension. For fractals, this is not always as easy, so let's formalize this procedure. One way to find the exponent in these three cases is to use the logarithm of the number of constituent pieces into which the object has been subdivided. For a line, we find

$$\log\,(\text{number of pieces}) = \log(n^1) = 1\log n$$

For a square,

$$\log\,(\text{number of pieces}) = \log(n^2) = 2\log n$$

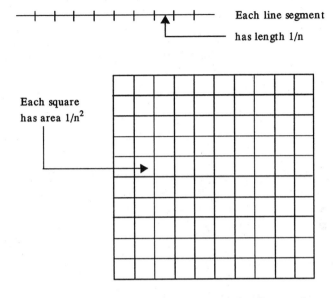

Each line segment has length 1/n

Each square has area 1/n²

Figure 9.13 Calculating the dimensions of a line and a square.

and for a cube,

$$\log(\text{number of pieces}) = \log(n^3) = 3\log n$$

Now remember that we divided the line, square, and cube into pieces which, when magnified by a factor of n, gave the original figure. So if we now divide log (number of pieces) by the logarithm of this magnification factor, we get the dimension. That is, the dimension D is given by the formula

$$D = \frac{\log(\text{number of pieces})}{\log(\text{magnification})}$$

So, for a line, we find

$$D = \frac{\log n^1}{\log n} = 1$$

For a square,

$$D = \frac{\log n^2}{\log n} = \frac{2\log n}{\log n} = 2$$

and for a cube,

$$D = \frac{\log n^3}{\log n} = \frac{3\log n}{\log n} = 3$$

Each of these calculations was easy because the magnification factor in each case was n. But what about our Sierpinski triangle? Recall that the lower left corner of this figure could be magnified by a factor of 2 to produce the whole triangle. On the other hand, the triangle consists of precisely 3 separate pieces, all identical to each other, namely, the lower left, the lower right, and the upper triangles. So mimicking what we did for the line, square, and cube, we find that the dimension of the Sierpinski triangle is

$$D = \frac{\log\,(\text{number of triangles})}{\log\,(\text{magnification})} = \frac{\log 3}{\log 2} = 1.585\ldots$$

which is by no means an integer! Let's try this again. The Sierpinski triangle may also be manufactured by assembling nine smaller pieces as we described in Section 9.2. Each of these smaller triangles is exactly one-fourth the size of the original figure. Hence

$$D = \frac{\log 9}{\log 4} = \frac{\log 3^2}{\log 2^2} = \frac{2\log 3}{2\log 2} = \frac{\log 3}{\log 2}$$

and we get the same answer.

To calculate the dimension of the Koch snowflake, we recall that each side of the original triangle is decomposed into four smaller pieces with a magnification factor of 3. Therefore,

$$D = \frac{\log 4}{\log 3} = 1.261\ldots$$

We use the sides of the snowflake because no piece of the snowflake may be magnified to look like the whole object; pieces of the sides are self-similar, however. If we proceed to the second stage of the construction, there are then 16 sides, but the magnification factor is 3^2. Again,

$$D = \frac{\log 4^2}{\log 3^2} = \frac{2\log 4}{2\log 3} = 1.261\ldots$$

Notice that the dimension of the snowflake is somewhat smaller than that of the Sierpinski triangle. This agrees with what our eyes are telling us. The triangle looks larger, more two-dimensional, than the snowflake curve, and hence it has a larger dimension.

Finally, for the Cantor set, the number of intervals at each stage of the construction is 2^n, but the magnification factor is 3^n. So,

$$D = \frac{\log 2^n}{\log 3^n} = \frac{n\log 2}{n\log 3} = .6039\ldots$$

Exercise 9.12 Compute the dimension of the middle-fifths Cantor set. Compute the dimension of the box-fractal in Figure 9.4.

Remark. The dimensions computed in this section were easy to compute because the magnification factor always increased at the same rate as the number of pieces in the figure. For many fractals, this is not the case. For example, the Julia sets of Chapters 6 and 7 often have fractional dimension, but this dimension is very difficult to compute. Indeed, for many Julia sets, the exact dimension is unknown. Nevertheless, a glance at Figure 7.5 tells us that these fractals should have different dimensions, and indeed they do.

9.6   Fractals and Dynamics

In this section we show how fractals arise naturally in dynamical systems. Of course, we have seen this to some extent already when we discussed Julia sets. Here we will show how the Cantor middle-thirds set arises naturally in a dynamical system.

Consider the function

$$T(x) = \begin{cases} 3x & x \le \frac{1}{2} \\ 3 - 3x & x \ge \frac{1}{2} \end{cases}$$

The graph of T is shown in Figure 9.14. T is called a *tent* function because of the shape of its graph. The following exercise shows that all the interesting orbits of T lie within the interval $0 \le x \le 1$.

Exercise 9.13 Use graphical analysis to show that if $x < 0$, then $T^n(x) \to -\infty$ as $n \to \infty$. Similarly, show that if $x > 1$, then $T^n(x) \to -\infty$ as $n \to \infty$.

So we need consider only orbits that remain forever in the interval $0 \le x \le 1$. As we have so often in the past, let's begin to investigate the dynamics by using the computer.

Experiment 9.14 Use ITERATE1 or ITERATE3 to compute various orbits of T with initial seed x_0 satisfying $0 \le x_0 \le 1$.

Outcome. You will probably find several orbits that do not leave $0 \le x \le 1$ under iteration. For example, the fixed points at $x_0 = 0$ and $x_0 = \frac{3}{4}$ do not leave. But it appears that almost all other orbits do leave.

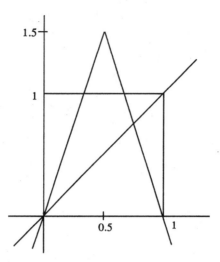

Figure 9.14 The graph of the tent function.

This should remind you of the situation with the logistic function $F(x) = cx(1 - x)$ for c-values larger than 4 (see Section 2.2). In fact, the situation in this case is very similar.

From the preceding exercise, we need consider only the orbits of points in $0 \leq x \leq 1$. However, many of these orbits eventually leave this interval. Indeed, Figure 9.15 shows that, if $\frac{1}{3} < x < \frac{2}{3}$, then $T(x) > 1$. This means that each point in $\frac{1}{3} < x < \frac{2}{3}$ leaves $0 \leq x \leq 1$ after one iteration. Hence we know what must happen: This orbit tends to $-\infty$. Thus we need consider only what happens to points in the remaining intervals, $0 \leq x \leq \frac{1}{3}$ and $\frac{2}{3} \leq x \leq 1$.

Here is where we begin to see a Cantor set emerging. Which points in $0 \leq x \leq \frac{1}{3}$ or $\frac{2}{3} \leq x \leq 1$ have orbits that remain for all iterations in $0 \leq x \leq 1$? Let's see. After one application of T, each of these intervals is stretched over the entire interval $0 \leq x \leq 1$, since, for example, $T(0) = 0$ and $T(\frac{1}{3}) = 1$. Hence there are points in this interval whose image lies in $\frac{1}{3} < x < \frac{2}{3}$. If x_0 is any such point, then $T^2(x_0) < 0$, since $T(x) > 1$ for any x in $\frac{1}{3} < x < \frac{2}{3}$. Therefore, as we have already seen, $T^n(x_0) \to -\infty$ as $n \to \infty$.

Clearly, if $\frac{1}{9} < x_0 < \frac{2}{9}$ or $\frac{7}{9} < x_0 < \frac{8}{9}$, then we have $\frac{1}{3} < T(x_0) < \frac{2}{3}$. Therefore, we may forget about these two intervals, because orbits of points here tend to $-\infty$. Figure 9.15 shows this.

Continuing in this fashion, we see that there are four intervals having the

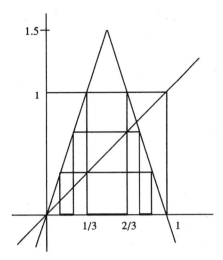

Figure 9.15 A Cantor set for the dynamics of T.

property that any point x_0 in them satisfies $T^3(x_0) > 1$. Therefore, these points have orbits tending to $-\infty$. Thus we see that we may disregard any middle-third interval, exactly as in the construction of the Cantor set. Stated in the opposite way, we see that it is precisely the Cantor middle-thirds set whose points have orbits that do not tend to $-\infty$. Thus, all the interesting dynamics for T take place on a fractal, the Cantor middle-thirds set.

Exercise 9.15 Sketch the graph of T^2, T^3, T^4 for $0 \le x \le 1$. Conclude that T^2 has 4 fixed points, T^3 has 8, and T^4 has 16. How many fixed points does T^n have?

Exercise 9.16. Let x_0 be any endpoint of the Cantor middle-thirds set. What can you say about the orbit of x_0? Conclude that none of the periodic points found in the previous exercise (except 0) are endpoints.

The previous two exercises confirm what we saw earlier — there are many more points in the Cantor set than the endpoints of the removed intervals. We can compute many of the periodic points of T. For example, $\frac{3}{10}$ and $\frac{9}{10}$ lie on a cycle of period 2, since $T(\frac{3}{10}) = \frac{9}{10}$ and $T(\frac{9}{10}) = \frac{3}{10}$. Note that the orbit of $\frac{3}{10}$ hops back and forth between the original left and right intervals in the Cantor set construction. We may therefore associate a sequence of lefts and rights to these points which give the "approximate" location of each of the points on this orbit. For example, the sequence associated to

$\frac{3}{10}$ is $LRLRLR\ldots$. Similarly, the sequence associated to $\frac{9}{10}$ is $RLRLRL\ldots$. Note that both of these sequences are repeating; this corresponds to the fact that the orbits themselves are periodic. Similarly, the point $\frac{1}{10}$ is eventually periodic, because $T(\frac{1}{10}) = \frac{3}{10}$; the associated sequence is $LLRLRLR\ldots$. One of the amazing facts about the Cantor set is that every point in the Cantor set has a unique representation as a sequence of Ls and Rs. Conversely, every sequence of Ls and Rs corresponds to a point in the Cantor set. This means that there are lots of different orbits for T in the Cantor set! This is also the beginning of a very interesting subject in mathematics known as symbolic dynamics.

Remark. The sequence of lefts and rights discussed above is different from the sequences discussed in Exercises 9.5 and 9.6. The earlier sequences are related to the ternary expansion of numbers in the Cantor set, while these new sequences are related to the dynamics of T.

Exercise 9.17 Verify that each of the following points is a periodic (or fixed) point for T and find its prime period.

 a. $x_0 = \frac{3}{4}$

 b. $x_0 = \frac{3}{10}$

 c. $x_0 = \frac{3}{28}$

 d. $x_0 = \frac{3}{82}$

What is the sequence of Ls and Rs associated to each of these points? Can you find a formula that gives a periodic point of period n for T? *Hint:* Do you see a pattern in the denominators of these points?

Exercise 9.18 Find a function that has the property that the set of points whose orbits do not escape to $-\infty$ is precisely the Cantor middle-fifths set.

Chapter 10

Chaos

Our goal in this section is to explain one of the most important recent discoveries in mathematics, the fact that very simple dynamical processes can behave in a very complicated, almost random fashion. We have seen this type of behavior already when we looked at the dynamics of the logistic function $F_c(x) = cx(1 - x)$, where we saw that, for many values of c, the successive points on the orbit of x under F_c seem to hop around the line unpredictably. We saw this with a number of our earlier programs, including ITERATE1, ITERATE2 and ORBITDGM. This is the concept of chaos, one of the most important new topics in mathematics.

Actually, chaos is all around us. From the swirling patterns of a hurricane on a meteorologist's radar scope to the eddies and swirls of a mountain stream, from the ups and downs of the stock market to the uncontrollable patterns formed by smoke as it rises, all these phenomena seem totally unpredictable and out of control. All these phenomena are inherently chaotic. You may protest that you know the reason why. All these examples — weather systems, the economy, hydrodynamic flows — involve countless variables. It would be physically impossible for any human being to understand and predict all of the facets of the economy or where each and every molecule of water will travel in a stream. This is certainly the case, but this is not necessarily what makes a system chaotic. As we will see, even simple functions like our quadratic family behave just as unpredictably when iterated. All the ingredients of chaos are present in this simple dynamical system.

This is the fundamental breakthrough made by mathematicians in recent years, the realization that chaotic systems need not depend on huge numbers of variables but may in fact depend on only one, as in the case of $cx(1 - x)$.

This discovery bodes well for scientists in all disciplines. The realization that chaos can be studied by elementary means will stimulate scientists to explore chaotic phenomena with a different approach, searching for the one or two variables that may cause a system to behave unpredictably. This in turn may lead to a new understanding of chaotic phenomena in nature, particularly those that previously seemed unfathomable.

As we mentioned in the introduction, our aim is to present only the mathematical ideas behind chaos. We will leave the applications of these ideas to others. You should remember, however, that these ideas are so new that many of them have yet to find applications in science and engineering. In the future, however, it seems plausible that these ideas will find many important applications.

10.1   The Squaring Function Again

The prototype of a chaotic function is the squaring function, $T(z) = z^2$, where z is complex. Of course, not all orbits of the squaring function are unpredictable. As we have seen, if $|z| < 1$, then the orbit of z tends to the attracting fixed point at 0. If $|z| > 1$, then the orbit of z tends to infinity. What remains are the points on the circle of radius 1, and it is these points whose orbits are chaotic. So our aim therefore becomes to explain the behavior of these orbits and why they are chaotic. Remember that the circle of radius 1 is precisely the Julia set. The behavior that we will uncover is typical of *all* Julia sets, not just the Julia set of z^2.

How do we understand the behavior of these orbits? Recall that a point on the circle of radius 1 can be written in polar form as

$$z = \cos\theta + i\sin\theta$$

Its image under the squaring function is given by

$$T(z) = z^2 = \cos^2\theta - \sin^2\theta + 2i\sin\theta\cos\theta$$
$$= \cos(2\theta) + i\sin(2\theta).$$

That is, the point with polar angle θ is moved by T to the point with polar angle 2θ.

We can use graphical analysis to understand how these polar angles change under iteration. Recall that the polar angle of a point is defined only up to a multiple of 2π. This means that the angles θ, $\theta + 2\pi$, $\theta + 4\pi$, ...

all represent the same point on the circle. We will always represent a point by its angle θ, where $0 \leq \theta < 2\pi$. This means that if we double an angle and get a result between 2π and 4π, we will simply subtract 2π from the result, thus yielding a polar angle between 0 and 2π. If you are familiar with modular arithmetic, we are simply multiplying angles by 2 mod 2π. We will denote this doubling function by D; that is,

$$D(\theta) = 2\theta \bmod 2\pi$$

Equivalently, for $0 \leq \theta < 2\pi$,

$$D(\theta) = \begin{cases} 2\theta & \text{if } 0 \leq \theta < \pi \\ 2\theta - 2\pi & \text{if } \pi \leq \theta < 2\pi \end{cases}$$

Note that if $0 \leq \theta < 2\pi$, $D(\theta)$ is less than 2π.

This formula allows us to use graphical analysis to understand the squaring function. The graph of D is shown in Figure 10.1. This function may be iterated using any of our programs from Chapter 2.

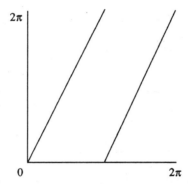

Figure 10.1 The graph of the doubling function.

Experiment 10.1 Use ITERATE1 or ITERATE2 to compute various orbits of D. Record what you find.

Outcome. For almost any $\theta > 0$, it appears that the orbit of θ "fills in" the entire interval $0 < \theta < 2\pi$.

This is by no means the case, as graphical analysis shows.

Exercise 10.2 Use graphical analysis to sketch the graphs of D^2, D^3, D^4,.....
How many fixed points do you find for D^2, D^3, D^4,...? How many fixed
points does D^n have?

This exercise shows that many, many points in the interval $0 \le x \le 2\pi$
have orbits that are periodic under D, but the computer fails to find them!
This happens because all of these periodic points are repelling and hence
unstable. We have seen this phenomenon before: Recall from Chapter 4
that the functions $x^2 - 2$ and $4x(1 - x)$ have similar properties.

10.2   Sensitive Dependence

Now we come to the essential ingredient of a chaotic system, *sensitive
dependence on initial conditions*. Suppose we take two points θ_0 and θ_1 on
the circle, and suppose that θ_0 and θ_1 are fairly close together. Suppose that d
is the distance measured along the circle between them. What happens when
we iterate T? Clearly, the distance between $T(\theta_0)$ and $T(\theta_1)$ has doubled (as
long as $T(\theta_0)$ and $T(\theta_1)$ are still close to each other). If we iterate again, the
distance between $T^2(\theta_0)$ and $T^2(\theta_1)$ doubles again. Continuing, we see that
the distance between $T^n(\theta_0)$ and $T^n(\theta_1)$ is $2^n d$. So the orbits of θ_0 and θ_1
are separating quite rapidly. After all, after only 10 iterations, the distance
between the two points is $2^{10}d = 1024d$. See Figure 10.2.

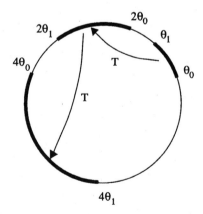

Figure 10.2 The squaring function doubles distances on the circle.

This means that, no matter how close two orbits start out, after just a few iterations they will be very far apart. The orbits separate from each other exponentially. Think of what this means in terms of the computer. We can specify a particular point on the circle only up to a finite number of decimal places; any more decimal places will simply be disregarded by the computer. This means that we will necessarily make a very small error when we input a typical number into the computer. Now begin iterating; very quickly we see that this small error is magnified so that the computed orbit is very far away from the actual orbit. A small error in initial conditions makes a big difference in what we see. This is sensitive dependence on initial conditions and is the typical behavior in a chaotic system.

Project 10.3 This project gives you another way to understand sensitive dependence on initial conditions. Modify the program ITERATE1 so that it displays not one orbit, but several orbits of the same function at the same time. Your output should be in several columns, each column containing the orbit of a different point. The first column should contain the iteration count. The resulting "spreadsheet" is useful for comparison of the behavior of orbits of nearby initial points. See Table 2.3 for a sample of the output from that program.

Experiment 10.4 Use the program in the previous project to compute and display the first 100 points on the orbits of the points $x_0 = .5$, $x_0 = .501$, and $x_0 = .5001$ for the function $F(x) = 4x(1 - x)$. What do you find?

Outcome. Despite the fact that the initial points .5, .501, and .5001 are all relatively close together, after only a very few iterations, their orbits bear no resemblance whatsoever to each other. This function possesses sensitive dependence on initial conditions.

Experiment 10.5 Try the previous experiment with a variety of different initial conditions. Record what you find. Do you ever find an initial condition in the interval $0 < x_0 < 1$ whose orbit is not sensitive to initial conditions? Remember that the computer uses only a finite amount of accuracy when reading your input, so if you use too many decimal places in your initial input, the computer will just forget the additional digits.

Experiment 10.6 Perform Experiment 10.4 for the function $Q(x) = x^2 - 2$ using initial points in the interval $-2 \leq x \leq 2$. Does this function depend sensitively on initial conditions?

There is more to the story of sensitive dependence when we consider Julia sets. To explain this, let's stay with the squaring function T. Suppose z_0 is a point on the unit circle, and suppose that in polar form z_0 is given by $r_0(\cos\theta_0 + i\sin\theta_0)$. Let's erect a small "chunk" of a wedge around z_0 by considering all complex numbers that satisfy $\theta_1 < \theta < \theta_2$ and $r_1 < r < r_2$, where $\theta_1 < \theta_0 < \theta_2$ and $r_1 < 1 < r_2$. This chunk is depicted in Figure 10.3. Notice that we can make this chunk as small as we like by picking r_1 and r_2 close to 1 and θ_1 and θ_2 close to θ_0. Let's call this chunk W.

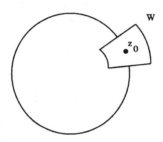

Figure 10.3

The amazing fact about the Julia set of T is that no matter how small we choose W, the images $T^n(W)$ eventually meet any point in the plane, with at most one exception. Figure 10.4 shows why this is true for the squaring function. As we iterate T, the chunk grows larger and larger: The outer radius increases toward infinity, whereas the inner radius shrinks to zero. Similarly the angle of the chunk is doubled at each stage. Thus we see that, no matter how small we choose W initially, iteration makes W expand until its image eventually hits any point in the plane (with one exception, namely, 0).

This means that points on the Julia set of T have orbits that depend very sensitively on initial conditions. Arbitrarily close to any point on the circle is another point whose orbit eventually hits any point whatsoever in the plane, with only one exception.

Amazingly, this property is true for all Julia sets, not just for the squaring function. We cannot prove this here, unfortunately. However, this fact gives one very good reason for the importance of the study of Julia sets. Points in the Julia set have orbits that are the most unstable of all orbits.

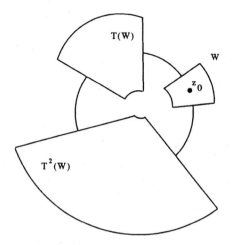

Figure 10.4

Further Exercises and Experiments

1. Consider the doubling function on the interval $0 \le x < 1$ given by

$$V(x) = \begin{cases} 2x & \text{if } 0 \le x < \frac{1}{2} \\ 2x - 1 & \text{if } \frac{1}{2} \le x < 1 \end{cases}$$

Use graphical analysis to decide how many periodic points of period n the function V has. All the remaining exercises in this section deal with the dynamics of V.

2. Compute the orbit under V of each of the following points:
   a. $x = \frac{1}{3}$
   b. $x = \frac{1}{5}$
   c. $x = \frac{1}{7}$
   d. $x = \frac{3}{7}$
   e. $x = \frac{1}{4}$
   f. $x = \frac{1}{16}$
   g. $x = \frac{9}{16}$

3. Suppose that x is a rational number of the form p/q, where p and q are integers and $0 \le p < q$. What can you say about the orbit of x under V if q is odd? What if $q = 2^n$ for some natural number n?

4. For each x satisfying $0 \leq x < 1$, let's associate a sequence of 0s and 1s to x that tells us roughly what happens to the orbit of x under V. This sequence of 0s and 1s will be called the itinerary of x. It is defined as follows. The first entry in the itinerary of x will be 0 if $x < \frac{1}{2}$ or 1 if $x \geq \frac{1}{2}$. The second entry will be 0 if $V(x) < \frac{1}{2}$, 1 if $V(x) \geq \frac{1}{2}$. In general, the kth entry will be 0 if $V^{k-1}(x) < \frac{1}{2}$ or 1 if $V^{k-1}(x) \geq \frac{1}{2}$. So the itinerary of the point 0 is 000... because 0 is fixed. The itinerary of $\frac{1}{2}$ is 1000.... Identify the itinerary of each of the points in Exercise 2.

5. Do you see anything peculiar about the itinerary of x? In particular, what is the relationship between the binary expansion of any x with $0 \leq x < 1$ and the itinerary of x?

6. What is the itinerary of $V(x)$, given the itinerary of x?

Chapter 11

Julia Sets of Other Functions

In previous chapters we have concentrated on Julia sets and the Mandelbrot set associated to the quadratic functions $Q_c(z) = z^2 + c$. There were several reasons for this, some historical and some mathematical. The quadratic family is the simplest nonlinear family of functions on the real line or complex plane exhibiting the rich mathematical structures that we have seen. But there are many, many other families of functions that exhibit these and other phenomena. In this chapter we sample some of the behavior of these functions. The beauty and complexity that we see is by no means limited to the special functions that we study here. We urge you to sample some of this interesting and appealing mathematics by experimenting with other functions.

11.1   Higher-Degree Polynomials

Although every polynomial has a Julia set, the easiest polynomials with which to work are those that have the special form $P_c(z) = z^n + c$ with $n = 3, 4, 5 \ldots$. The exponent n is called the degree of this polynomial. Like the quadratic functions, these polynomials have a single critical point at 0, since 0 is the only point whose image is c. To compute the Julia set of P_c, you may use any of the three methods discussed in Chapters 6 and 7, but some modifications are necessary. The first difficulty that occurs is determining when an orbit escapes to infinity under P_c. The criterion for this turns out to be exactly the same as in the case of quadratic functions. You should verify this yourself by mimicking what we did in previous sections for Q_c.

Exercise 11.1 For the family of functions $P_c(z) = z^n + c$, show that:

a. If $|z| \geq |c|$ and $|z| > 2$, then $|P_c(z)| > |z|^{n-1}(1 + \ell)$, where $\ell > 1$.

b. Conclude from part (a) that if $|c| > 2$, then the critical orbit of P_c tends to infinity.

c. Conclude also that if $|c| \leq 2$, the filled in Julia set of P_c lies inside the circle of radius 2 centered at the origin.

This exercise allows us to modify our earlier Julia set programs so that they compute the Julia sets for $P_c(z) = z^n + c$.

Project 11.2 Modify JULIA1 and JULIA3 so that they compute and display the Julia sets of $P_c(z) = z^n + c$. Your programs should allow the user to input both n and c.

Experiment 11.3 Use these Julia set programs to investigate the dynamics of the functions $T_c(z) = z^3 + c$ for real values of c. What do you observe? Is there any period doubling? Can you find saddle-node bifurcations? Can you explain this using graphical analysis of the real function $T_c(x) = x^3 + c$?

Outcome. It appears that either the Julia set is totally disconnected or else consists of one large region. Graphical analysis shows that either T_c has a single attracting fixed point or else the orbit of the critical point 0 escapes to infinity. In Figure 11.1, we examine a bifurcation near $c = .3849\ldots$. Can you explain this?

Experiment 11.4 Use these modified programs to check that the polynomials P_c have basically the same properties as the quadratic functions. For example,

a. If the critical orbit escapes, the Julia set is fractal dust.

b. If P_c has an attracting cycle, then the critical orbit is attracted to it (use ITERATE4).

c. If P_c has an attracting cycle, then the Julia set of P_c is connected.

The backward iteration method for computing Julia sets works for P_c as well, but the algorithm is more complicated. To use this method, we must compute nth roots rather than square roots. Each nonzero complex number has exactly n of these nth roots. This can be seen by using the polar representation of a complex number introduced in Chapter 6.

Suppose

$$z = r(\cos\theta + i\sin\theta)$$

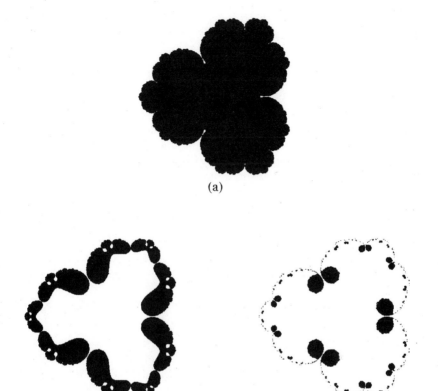

(a)

(b)   (c)

Figure 11.1 The Julia sets of (a) $z^3 + .384$, (b) $z^3 + .3883$ and (c) $z^3 + .3885$ computed using JULIA1 with 50 iterations.

Then we know that

$$z^2 = r^2(\cos 2\theta + i \sin 2\theta)$$

Therefore we may use complex multiplication to find

$$
\begin{aligned}
z^3 &= z \cdot z^2 \\
&= r^3((\cos\theta\cos 2\theta - \sin\theta\sin 2\theta) + i(\cos\theta\sin 2\theta + \sin\theta\cos 2\theta)) \\
&= r^3(\cos(\theta + 2\theta) + i\sin(\theta + 2\theta)) \\
&= r^3(\cos 3\theta + i\sin 3\theta)
\end{aligned}
$$

Using this fact and complex multiplication again, we find

$$z^4 = r^4(\cos 4\theta + i \sin 4\theta)$$

and, in general,

$$z^n = r^n(\cos n\theta + i \sin n\theta)$$

That is, raising z to the n th power is the same as raising its modulus to the n th power and multiplying its polar angle by n.

So how do we use this to find n th roots? Recall that, for square roots, we simply took the square root of the modulus and then selected either $\theta/2$ or $\theta/2 + \pi$ for the polar angle. For n th roots, the procedure is similar but more complicated. If

$$z = r(\cos \theta + i \sin \theta)$$

then the n th root of z has modulus $r^{1/n}$ and polar angle one of

$$\frac{\theta}{n}, \frac{\theta + 2\pi}{n}, \frac{\theta + 4\pi}{n}, \dots, \frac{\theta + 2(n-1)\pi}{n}$$

For example, the cube roots of 1 all have modulus 1. The polar angles are

$$\theta = 0, \frac{2\pi}{3}, \frac{4\pi}{3}$$

which we find by setting $\theta = 0$ and $n = 3$ in the preceding list. It follows that the cube roots of 1 are

$$1 = \cos 0 + i \sin 0$$

$$-\frac{1}{2} + \frac{\sqrt{3}}{2} i = \cos(2\pi/3) + i \sin(2\pi/3)$$

$$-\frac{1}{2} - \frac{\sqrt{3}}{2} i = \cos(4\pi/3) + i \sin(4\pi/3)$$

You may check this easily by simply cubing all of these complex numbers. Similarly, the fourth roots of $16i$ all have modulus 2. The polar angles are

$$\theta = \frac{\pi}{8}, \frac{5\pi}{8}, \frac{9\pi}{8}, \frac{13\pi}{8}$$

Exercise 11.5 Compute the n th roots of each of the following complex numbers:

a. $z = -1, n = 3$
b. $z = 1, n = 4$
c. $z = 8i, n = 3$
d. $z = 2, n = 6$

Project 11.6 Use these facts presented in the section to modify JULIA2 so that the new program computes the Julia set of P_c using the backward iteration method. The major change from our previous JULIA2 will be the random selection of one of the nth roots. How will you accomplish this? Use this program to experiment with various Julia sets of higher-degree polynomials.

Since the polynomials P_c have a single critical orbit, there is an analogue of the Mandelbrot set for each integer n. As for the quadratic functions, this set is the collection of c-values for which the critical orbit does not escape. By our experiment above, this is precisely the set of c-values for which the Julia set is connected. This set is called the *degree* n *bifurcation set*.

Project 11.7 Modify MANDELBROT1–4 so that the new programs display the degree n bifurcation set. Some of the results of these modifications are displayed in Figure 11.2.

Figure 11.2 The degree 3 and 4 bifurcation sets.

The polynomials $P_c(z) = z^n + c$ are quite special because they have only one critical orbit. Most polynomials have more than one of these orbits. The computation of critical orbits in more general cases necessitates the use of calculus (critical points are simply the points at which the derivative vanishes), so we won't attempt to go into depth on this matter here. For example, using elementary calculus, it can be shown that the cubic polynomial

$$P(z) = z^3 + Az + B$$

has two critical points given by $\pm\sqrt{-A/3}$.

For general polynomials, the algorithms used to produce Julia sets before still work, since the Julia set is the boundary between the orbits that escape and those that do not. Therefore, you may easily modify JULIA1 or JULIA3 to work for a general polynomial. The backward iteration method is not practical however because it is generally impossible to solve an equation such as

$$z^n + a_{n-1}z^{n-1} + \cdots + a_1 z + a_0 = w$$

for z. This would be the essential step in backward iteration.

Project 11.8 Use JULIA2 and JULIA3 to compute various Julia sets of degree n polynomials. Some typical patterns are displayed in Figure 11.3.

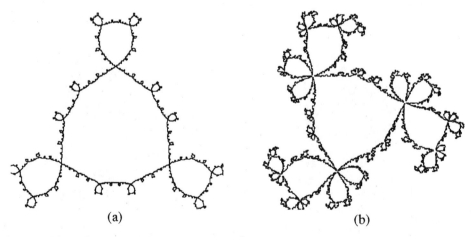

(a)   (b)

Figure 11.3 Julia sets for (a) $z^3 + i$ and (b) $z^3 - .574 + .271i$
computed using JULIA3.

11.2   Euler's Formula

In this section we introduce one of the most interesting and surprising formulas in all of mathematics, Euler's formula. We will need this formula in the next section when we investigate the Julia sets of transcendental functions.

Euler's formula relates the exponential function to the trigonometric functions $\sin x$ and $\cos x$. The usual exponential function is denoted by e^x or $\exp x$, where e is the base of the natural logarithm. The number e is approximately equal to 2.7128.... This function is important in many areas of mathematics and science, where it is used to measure growth and decline processes such as population growth, compound interest, and radioactive decay. The graph of e^x is displayed in Figure 11.4.

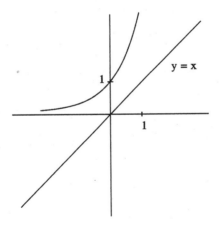

Figure 11.4 The graph of the exponential function.

Euler's formula allows us to begin discussing complex analogues of the exponential and trigonometric functions. These functions are not as well known as their real counterparts, but we will see that their Julia sets are quite spectacular from a geometric and dynamic point of view. Euler's formula is

$$e^{ix} = \cos x + i \sin x$$

We cannot prove this formula here. You will see why it is true if you study infinite series as part of a calculus course. We ask you at this stage simply to accept this formula as the definition of the exponential of an imaginary number. Note the surprising fact that the exponential function, which intuitively has nothing whatsoever to do with trigonometry, is actually a sum of sines and cosines. This formula has many ramifications in mathematics, in particular in the study of differential equations, a field that is one of the most important applications of calculus.

Letting $x = \pi$ in Euler's formula yields the amazing fact that

$$e^{i\pi} = \cos \pi + i \sin \pi = -1$$

There are perhaps no "stranger" numbers in all of mathematics than e, i, and π. Yet when these three numbers are combined as above, they give the simple value -1!

Euler's formula allows us to define the complex exponential function. By the usual rules of exponentiation, we should have

$$e^{x+iy} = e^x \, e^{iy}$$

By Euler's formula, if $z = x + iy$, we must therefore have

$$e^z = e^{x+iy} = e^x(\cos y + i \sin y)$$

This is the formula that allows us to use the real exponential, sine, and cosine to compute the complex exponential function.

Exercise 11.9 Compute the value of e^z for each of the following:
    a. $z = 2 + i\pi$
    b. $z = 2\pi i$
    c. $z = 1 + i\pi/4$
    d. $z = 4 = 4 + 0i$
    e. $z = \pi + 2i$

One major difference between the exponential function and polynomials is the fact that points that are far from the origin in the complex plane are no longer taken even further away by the function. For example, if $z = -10$, then

$$e^z = e^{-10} = .000045\ldots$$

which is very close to 0. This fact will become important when we discuss the Julia set of the exponential in the next section.

11.3   Julia Sets of Transcendental Functions

Functions such as the exponential, the sine, and the cosine are called transcendental functions. Unlike polynomials, they cannot be computed by a finite number of arithmetic operations. Nevertheless, using the computer, we may still evaluate and iterate these functions.

For transcendental functions, we concentrate on the set of points whose orbits escape. For technical reasons, this set contains the Julia set. This may appear to contradict the definition we gave for polynomials (the Julia set was the *boundary* of the set of escaping points for polynomials), but actually it does not. For transcendental functions such as the sine or cosine, any point whose orbit escapes is actually on the boundary of the set of escaping orbits. That is, arbitrarily close to any point whose orbit escapes there must be a point whose orbit does not (a repelling periodic point, for instance). This means that all escaping points lie in the Julia set.

For transcendental functions, orbits usually escape to infinity in certain preferred directions that depend upon the function. For the complex exponential, any orbit that escapes must do so with an ever-increasing real part. To see why this is so, consider a point on the negative real axis such as -10. The image of this point, as we saw earlier, is $e^{-10} = .000045\ldots$, which is very close to 0. Similarly, if we take a point of the form $-10 + iy$, the image is

$$e^{-10}(\cos y + i \sin y)$$

The modulus of this point is again e^{-10}. (Do you see why?) So the image of $-10 + iy$ is again close to 0, no matter what y is. On the other hand, e^{10} is approximately equal to $22,026$, which is quite large, and

$$e^{10+iy} = e^{10}(\cos y + i \sin y)$$

which has modulus e^{10}.

Thus to compute the Julia set of the exponential, we will make use of the following algorithm:

1. Select a 200×200 grid in the plane.
2. Iterate each point in the grid up to 20 times.
3. If the real part of any point on the orbit is larger than 50, stop the iteration and color the original point white.

4. Otherwise, color the original point black. Thus, this algorithm produces a picture whose white points lie in the Julia set.

As part of this algorithm, we use the bound 50 to check whether an orbit escapes. This may seem like a relatively small number, but think of the image of 50 under the exponential; $e^{50} > 5 \times 10^{21}$ is huge, and the next image will be astronomically large!

Project 11.10 Write a program called JULIAEXP that uses the preceding algorithm to compute the Julia set of $E_c(z) = ce^z$, where c is a complex parameter.

Remark. On many computers, if you encounter a small number such as e^{-50}, the computer will round off this number and give you 0 as a result. This is perfectly acceptable in this program. On some computers, however, you may receive an "underflow" error message. If this is the case, you may remedy this by simply redefining $e^{x+iy} = 0$ whenever $x \leq -50$. This will have the effect of eliminating the underflow.

Experiment 11.11 Use JULIAEXP to compute the Julia sets of ce^z for the following c-values:

    a. $c = .2$
    b. $c = 1$
    c. $c = -4 + i$
    d. $c = 1 + 2i$.

You should compute these Julia sets in the square $-3 \leq x \leq 3$, $-3 \leq y \leq 3$, although other windows may show more detail.

Outcome. Note that the Julia sets for exponential functions look quite different from those of polynomials. See Figure 11.5.

As you might expect from Euler's formula, there is a complex analogue of the sine and cosine. To derive formulas for these functions, we begin with Euler's formula for x and $-x$:

$$e^{ix} = \cos x + i \sin x$$
$$e^{i(-x)} = \cos(-x) + i \sin(-x)$$
$$= \cos x - i \sin x$$

This last equality follows from the facts from trigonometry that $\cos(-x) = \cos x$ and $\sin(-x) = -\sin x$. If we now add these two equations and multiply

Figure 11.5 The Julia set of $(1 + 2i)e^z$.

the result by $\frac{1}{2}$, we find

$$\frac{1}{2}\left(e^{ix} + e^{-ix}\right) = \cos x$$

Subtracting and multiplying by $\frac{1}{2}$, we find

$$\frac{1}{2}\left(e^{ix} - e^{-ix}\right) = i \sin x$$

Since $-i \cdot i = 1$, multiplication by $-i$ thus yields

$$\frac{-i}{2}\left(e^{ix} - e^{-ix}\right) = \sin x$$

This motivates our definition of the complex sine and cosine as

$$\sin z = -\frac{i}{2}\left(e^{iz} - e^{-iz}\right)$$
$$\cos z = \frac{1}{2}\left(e^{iz} + e^{-iz}\right)$$

If you have encountered the hyperbolic sine and cosine functions in your mathematical travels, you will note a close relationship between them and the complex sine and cosine. Indeed, if $z = x + iy$, we may write

$$\sin z = \sin(x + iy)$$

$$= -\frac{i}{2}\left(e^{i(x+iy)} - e^{-i(x+iy)}\right)$$

$$= -\frac{i}{2}\left(e^{-y}e^{ix} - e^{y}e^{-ix}\right)$$

$$= -\frac{i}{2}\left(e^{-y}(\cos x + i\sin x) - e^{y}(\cos(-x) + i\sin(-x))\right)$$

$$= -\frac{i}{2}\left(e^{-y}\cos x + ie^{-y}\sin x - e^{y}\cos x + ie^{y}\sin x\right)$$

$$= \left(\frac{e^{y} + e^{-y}}{2}\right)\sin x + i\left(\frac{e^{y} - e^{-y}}{2}\right)\cos x$$

The expressions

$$\frac{e^{y} + e^{-y}}{2}, \quad \frac{e^{y} - e^{-y}}{2}$$

have names; they are the *hyperbolic cosine* and *hyperbolic sine functions*, defined by

$$\cosh y = \frac{e^{y} + e^{-y}}{2}$$

$$\sinh y = \frac{e^{y} - e^{-y}}{2}$$

These functions are not always available as built-in functions in BASIC. However, they are easily incorporated into BASIC programs using the function definition DEF FN statement.

Thus we have

$$\sin(x + iy) = \sin x \cosh y + i\cos x \sinh y$$

which is reminiscent of the trigonometric formula for the sine of a sum of two numbers. The analogous formula

$$\cos(x + iy) = \cos x \cosh y - i\sin x \sinh y$$

also holds; you should check this for yourself.

These expressions allow us to compute the complex sine and cosine functions in terms of real exponential, sine, and cosine functions. For example,

$$\sin(i\pi) = \frac{-i\left(e^{-\pi} - e^{\pi}\right)}{2}$$

$$\cos(i\pi) = \frac{\left(e^{-\pi} + e^{\pi}\right)}{2}$$

Exercise 11.12 Evaluate each of the following expressions:

   a. $\cos(\pi + i\pi)$

   b. $\sin(\pi/2 + i\pi)$

   c. $\cos(i)$

   d. $\cos(2\pi)$ (Remember, $e^{\pi i} = -1$, so $e^{2\pi i} = e^{\pi i + \pi i} = 1$.)

   e. $\sin(2\pi + 4i)$

As with the complex exponential, points whose orbits escape to infinity lie in the Julia sets of sine and cosine. For sine and cosine, the direction of approach to infinity is different from the exponential. Orbits tend to infinity in the direction of either the positive or negative imaginary axis. For this reason, we say that an orbit "escapes" if its imaginary part ever becomes larger than 50 in absolute value.

Figure 11.6 displays a program called JULIASIN, which computes the Julia set of $\sin z$. As in the program JULIAEXP, this program displays the Julia set in white, not black. The image generated by this program, the Julia set of $\sin z$, is shown in Figure 11.7. Note that all points on the real axis are colored black, meaning that they are not in the Julia set — a fact we know from Chapter 1 where we saw that all orbits of $\sin x$ for real values of x tended to 0.

Project 11.13 Modify JULIASIN so that it computes

   a. The Julia set of $c \sin z$ for complex c.

   b. Enlargements of the Julia sets of $c \sin z$.

   c. The Julia set of $c \cos z$ via a program called JULIACOS.

The programs in this project take quite a while to run because of the many evaluations of the sine, cosine, and exponential functions necessary to compute the orbits. These function calls are much more time-consuming than the simple additions and multiplications necessary to compute polynomials.

Experiment 11.14 Use JULIASIN or JULIACOS to compute the Julia sets of

   a. $\cos z$

   b. $.5i \cos z$

   c. $i \sin z$

   d. $2 \sin z$

   e. $\pi \cos z$

   f. $2.98 \cos z$

```
REM  program  JULIASIN
FOR  i = 0 TO 200
     FOR  j = 0  TO  200
          x0 = -4 + 4 * i / 100
          y0 = 4 - 4 * j / 100
          x = x0
          y = y0
          FOR  n = 1 TO 25
               x1 = SIN (x) * ( EXP (y) + EXP (-y)) / 2
               y1 = COS (x) * ( EXP (y) - EXP (-y)) / 2
               IF  ABS (y1) > 50 GOTO 20
               x = x1
               y = y1
               NEXT  n
          PSET  (i, j)
20   NEXT j
NEXT  i
END
```

Figure 11.6 The program JULIASIN.

g. $2.96 \cos z$

In each case, it is best to begin with the square $|x|, |y| \leq 4$.

Outcome. For $\cos z$, recall that all points on the real axis tend to the fixed point $.73908\ldots$ under iteration of the cosine function. The black region found in Figure 11.8 represents the basin of attraction of this fixed point in the complex plane. See also Figure 11.9.

11.4   Exploding Julia Sets

Recall that the Julia sets of polynomials occasionally underwent dramatic changes when the function experienced a saddle-node or period-doubling

Figure 11.7 The Julia set of sin z.

Figure 11.8 The Julia set of cos z.

bifurcation. The same is true for transcendental functions. Often, these changes are quite spectacular for the complex sine, cosine, or exponential. The following experiments allow you to analyze some of these bifurcations, which we call *explosions*. This complex dynamical behavior was observed for the first time in the mid-1980s.

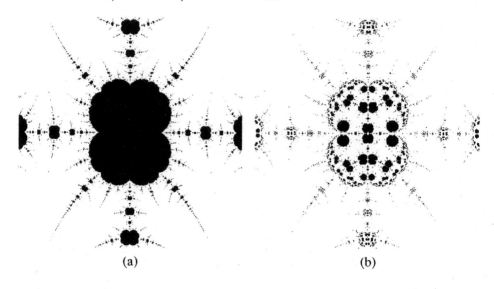

(a)   (b)

Figure 11.9 The Julia sets of $2.98 \cos z$ and $2.967 \cos z$.

Experiment 11.15 Consider the family ce^x with $c > 0$. Sketch the graph of ce^x for various c-values. For which c-values do you expect to find an attracting fixed point? For which c-values do you expect all real numbers to have orbits which escape? Now compute the Julia sets for the complex exponential ce^z with $c > 0$. What do you see? Record your observations.

Outcome. For small, positive values of c, the graph of ce^x crosses the diagonal in two places, whereas if c is large enough, the graph never crosses the diagonal. The precise point at which the graph changes its configuration is $c = 1/e = .36788\ldots$, where e is the base of the natural logarithm. The Julia sets of ce^z change dramatically at $c = 1/e$. In fact it is known that the Julia set is quite small when $c < 1/e$ and that the Julia set is the whole complex plane if $c > 1/e$. This is quite a dramatic change indeed!

Experiment 11.16 A similar phenomenon occurs in the family $ic \cos z$ for $c > 0$. Can you determine where this occurs? Can you explain this?

Experiment 11.17 Compute the Julia sets of $(1 + ci) \sin z$ for $c \geq 0$. When $c = 0$, we find the Julia set of $\sin z$, as depicted in Figure 11.6. When c increases, these Julia sets begin to change in an interesting fashion. However, nobody can explain these changes fully.

Further Exercises and Experiments

1. Consider the family of functions $F(z) = z^3 + ic$ where c is a real parameter.
   a. Show that F preserves the imaginary axis in the sense that $F(iy) = i(c - y^3)$.
   b. Use ITERATE1 or ITERATE2 to experiment with the real function $h(y) = c - y^3$. For which values of c does this function undergo a period-doubling bifurcation? Are there any subsequent period doublings?
   c. Use JULIA1 or JULIA2 to plot the Julia sets of these functions for various c-values.
   d. For which c-values does 0 lie on a cycle of period 2?
   e. Use JULIA2 to determine experimentally which values of c lead to Julia sets that resemble fractal dust.
   f. Can you explain all of this using graphical analysis?
   g. What does the degree 3 bifurcation set predict about this family and its period doublings?

2. Compute the Julia set of $(.6 + .8i) \sin z$ and $(.61 + 81i) \sin z$. Use 200 iterations to determine whether or not a point escapes in the latter case. Do you see another explosion?

3. The bifurcation set for the family $c \exp z$ may also be computed. All we need to know is which orbits play the role of the critical orbit, as in the case of the Mandelbrot set. For exponential functions, the critical orbit is always the orbit of 0. The reason for this is that 0 is the omitted value of $\exp z$: There is no z for which $\exp z = 0$. The algorithm is then to test each point in a grid in the complex c-plane to see if the corresponding critical orbit of $c \exp z$ ever has a real part exceeding 50. If so, color c black; if not, leave c white. Write a program called EXPBIF that effects this algorithm and then use it to compute the bifurcation set of the exponential function.

For Further Reading

There are a number of books that have appeared in recent years and that shed light on the topics of chaos, fractals, and dynamical systems from a variety of points of view. We particularly recommend the following.

Gleick, J. *Chaos: Making a New Science.* New York: Viking, 1987.
This book is the story of how chaos and fractals were found to play such an important role in all areas of science and mathematics in the 1970s and 1980s. It is basically a series of portraits of the scientists who discovered these ideas and how they worked in isolation at first. It is a wonderful and readable account of how these mathematical ideas developed.

Peitgen, H.-O., and Richter, P. *The Beauty of Fractals.* New York: Springer-Verlag, 1986.
This "coffee table" book features many beautiful color illustrations of Julia sets, the Mandelbrot set, and other images from dynamical systems theory. The text itself is a more advanced treatment of some of the topics we have discussed in this book. There are also a number of references to the scientific research literature.

Peitgen, H.-O., and Saupe, D., eds. *The Science of Fractal Images.* New York: Springer-Verlag, 1988.
This book presents a series of five articles, which give a "how-to" approach to using the computer to generate the fractal images from dynamical systems theory. Many of the algorithms presented in this book are refinements and extensions of the algorithms presented in this book. Consequently, the level of the presentation is more advanced.

Mandelbrot, B. *The Fractal Geometry of Nature.* San Francisco: Freeman, 1983.

This book is the author's definitive statement on the scope and purpose of fractal geometry. At times the book may be difficult to read, even for experts, but it certainly contains a wealth of ideas and applications.

Barnsley, M. *Fractals Everywhere*. Boston: Academic Press, 1988.

This is one of the first textbooks on fractal geometry. It is more advanced in the sense that readers should be familiar with both calculus and linear algebra. It contains a number of applications of the subject in the area of data and image compression.

Devaney, Robert L. *An Introduction to Chaotic Dynamical Systems*, 2nd ed. Menlo Park: Addison-Wesley, 1989.

We modestly recommend this book to readers who have a background in calculus and who wish to pursue the study of the mathematical aspects of chaotic dynamical systems theory.

Index

Asymptotic, 36, 71
Attracting fixed point, 45-48
Attracting periodic point, 51, 52

Backward iteration, 102, 160
Backward orbit, 101
Barnsley, M., 143
Basin boundary, 86
Basin of attraction, 46, 86
Bifurcation, 59, 63
Bifurcation theory, 31
Boundary-scanning method, 111, 123

Cantor set, 107, 133-136
Changing coordinates, 33, 80
Chaos, 68, 151-158
CIRCLE, 82
CLS, 35
Complex number, 75, 76

Complex conjugate, 118
Connected set, 108
Cosine (complex), 168, 169
Critical point, 110, 113, 159
Critical orbit, 110, 113
Cycle, 17

DEF FN, 170
Degree, 159
Degree n bifurcation set, 163
Diagonal, 40
Douady, A., 123
Douady's rabbit, 3
Dynamical systems, iii

Euler's formula, 165
Eventually fixed, 18
Eventually periodic, 18
Explosions, 173
Exponential function, 165

Filled-in Julia set, 90-96
Fixed point, 13, 17
FOR-NEXT, 24
Fractal, 3, 93, 129-150
FRACTAL, 142
Fractal dust, 95, 97, 107-111
Fractional dimension, 130,
    144-147
Function, 7
Functional notation, 10-12

GOTO, 70
Graph, 39
Graphical analysis, 39, 40-43

Higher iterates, 52-54
Hubbard, J., 123
Hyperbolic functions, 170

IF, 71
IF-THEN, 89
INPUT, 24
Invariant, 82
Imaginary part, 75
INT, 104
Iterated function system, 143
ITERATE1, 24
ITERATE2, 35
ITERATE3, 37
ITERATE4, 78, 79
ITERATE5, 83
Iteration, 2

Julia set, 3, 81-84, 85-112, 159
JULIAEXP, 168
JULIACOS, 171
JULIASIN, 171
JULIA1, 88
JULIA2, 103

Koch snowflake, 137-139

Logistic equation, 19
Logistic function, 22, 26-31, 128

Mandelbrot, B., 115
Mandelbrot set, 4, 113-128
MANDELBROT1, 117
MANDELBROT2-4, 121-124
Modulus, 76, 98
Multiplication (complex), 76

NEXT, 24
n th root, 162

Operation, 7, 8
Orbit, 13
Orbit analysis, 15
Orbit diagram, 70, 127
ORBITDGM, 70, 128

Parameter, 28
Period-doubling bifurcation, 63, 160
Periodic orbit, 17
Polar angle, 98
Polar representation, 97, 160
Prime period, 17
PRINT, 24
PSET, 35, 80

Quadratic family, 57

Random orbit, 141
Real part, 75
Removals, 130, 133
Repelling fixed point, 45, 48, 49
Repelling periodic point, 51, 52
RND, 104

Saddle-node bifurcation, 59, 160

Self-similarity, 93, 130
Sensitive dependence, 154-156
Sierpinski triangle, 130-132
Sine (complex), 168, 169
Square root (complex), 99
SQR, 103
Stable, 50
STEP, 70
Symbolic dynamics, 149

Tangent bifurcation, 59
Tent function, 147
Totally disconnected, 107
Transcendental function, 167
Triangle inequality, 77, 86, 115

Unstable, 50